red cell metabolism

A MANUAL OF BIOCHEMICAL METHODS
second edition

also by Dr. Beutler

Mechanisms of Anemia (*with I. M. Weinstein*), 1962
Clinical Disorders of Iron Metabolism, *first edition* (*with V. F. Fairbanks and J. L. Fahey*), 1963
Hereditary Disorders of Erythrocyte Metabolism, 1968
Clinical Disorders of Iron Metabolism, *second edition* (*with V. F. Fairbanks and J. L. Fahey*), 1971
Hematology (*with W. H. Williams, A. Erslev, and R. W. Rundles*), 1972
Symposium on Red Blood Cell and the Lens, (*with J. H. Kinoshita*), 1971
Red Cell Metabolism: A Manual of Biochemical Methods, *first edition,* 1971

red cell metabolism

A MANUAL OF BIOCHEMICAL METHODS
second edition

ERNEST BEUTLER, M.D.

*Chairman, Division of Medicine, and Director,
Department of Hematology, City of Hope Medical
Center, Duarte, California; and Clinical Professor
of Medicine, University of Southern California,
Los Angeles, California*

GRUNE & STRATTON
New York   San Francisco   London

A SUBSIDIARY OF HARCOURT BRACE JOVANOVICH, PUBLISHERS

Library of Congress Cataloging in Publication Data

Beutler, Ernest.
  Red cell metabolism.

  Includes bibliographical references and index.
  1. Enzymes—Analysis. 2. Erythrocytes.
I. Title. [DNLM: 1. Erythrocytes—Metabolism—
Laboratory manuals. WH25 B569r]
QP96.B45 1975   612′.111   74-32486
ISBN 0-8089-0861-8

Grune & Stratton, Inc.
111 Fifth Avenue
New York, New York 10003

Library of Congress Catalog Card Number 74-32486
International Standard Book Number 0-8089-0861-8
Printed in the United States of America

to Bonnie

contents

preface to the second edition

Since the publication of the first edition of this manual in 1971, the methods which it contained have been used extensively, not only in our own laboratory but in the laboratories of many of our colleagues. While the techniques described in the first edition have been very satisfactory, many improvements have become possible as greater experience in their use has accumulated. For example, the crucial importance of removing white cells and platelets more completely than is achieved merely by washing erythrocytes has become more apparent. Thus in the revised procedures presented in this edition, we recommend removal of leukocytes and platelets by filtration either through cotton-wool or through a mixture of microcrystalline and α cellulose. Further experience has taught us that it is feasible to carry out virtually all of the red cell enzyme assays on a freeze-thaw hemolyzate, rather than using three different hemolyzing solutions, as was proposed in the first edition. In reevaluating some of the assay procedures, it was found to be advantageous to make certain changes in the concentration of some of the substrates. Furthermore, we now propose testing the allosteric properties of two of the glycolytic enzymes, pyruvate kinase and phosphofructokinase, with limiting concentrations of fructose diphosphate and ADP, respectively. I am particularly grateful to Dr. Karl Blume of the University of Freiburg, West Germany, whose collaboration in reevaluating and modifying procedures for the assay of glycolytic enzymes was invaluable.

Since the first edition was prepared, several new techniques have been standardized in our laboratory. These include assays for the enzymes acetylcholinesterase, catalase, and diphosphoglyceromutase. These assays and fluorometric procedures for the estimation of glycolytic intermediates in erythrocytes are all included in the present edition. The methods for NAD + NADH and for NADP + NADPH have been deleted from the manual, because we have found that although recoveries of added pyridine nucleotides are very satisfactory in these techniques, extraction of endogenous pyridine nucleotides was incomplete.

Certain aspects of the methods for calculation of results of enzyme assays proved to be difficult to understand for a number of readers. Therefore some of these have been altered and, hopefully, simplified. The presentation of some of the material has also been reorganized to make the manual easier to use. For example, all normal values are summarized in a single table in the Appendix. With the advances in computer technology, the use of computer programs written for a single desk-top computer no longer seems appropriate or useful, so that the programs written for the Olivetti Underwood Programma 101 have been deleted from this edition.

It is again impossible for me to acknowledge specifically all the contributions of those who have made this revised manual possible, but I feel very keenly my debt to Dr. Karl Blume who devised many of the revised methods and helped to standardize the techniques. Most of the fluorometric methods for the assaying of red cell intermediates were developed together with Dr. Herwig Niessner. Mr. Steven Beutler rendered valuable editorial assistance. I am also particularly grateful for the continued technical assistance of Miss Florinda Trinidad, now Mrs. Florinda Matsumoto, and for the secretarial assistance of Miss Janet Muller, now Mrs. Janet Manning, who helped prepare the first edition.

The development of the techniques described has been supported in part by U.S. Public Health Service Grants HE 07449 from the National Heart Institute, and HD 01974 from the Heart and Lung Institute.

part I

introduction

1

purpose and organization
of the manual

In the past twenty years, there has been a tremendous increase in our knowledge of red blood cell metabolism. Previously unexplained hemolytic disorders are now known to be due to hereditary enzyme deficiencies. Intermediates in red cell metabolism, 2,3-diphosphoglycerate and ATP, have been shown to be regulators for the delivery of oxygen from the red cell to the tissues. Metabolic diseases can be detected by examination of red cells. The clinician, the physiologist, the geneticist, and the biochemist have all found the red cell to be a valuable tool in the expansion of knowledge in their disciplines. An increasing number of laboratories have therefore found it desirable to measure the activity of red cell enzymes or metabolic intermediates. Their task has not always been an easy one, since in many instances methods for the study of the particular substance in the red blood cell have not been published. Often, methods have been hastily adapted to study of the erythrocyte, sometimes with unfortunate results. From a practical point of view, another annoying difficulty often presents itself. Each technique gleaned from the literature requires the use of a different buffer for each enzyme assay, a different concentration of coenzymes and cofactors, and often a different final assay volume. Thus performing several enzyme assays involves the vexing task of preparing many solutions where a few would suffice, preparing a different hemolyzate for each assay, and setting up different calculations after each determination has been carried out. Often normal values are not available, and, even if control determinations are carried out, errors in the procedure may pass unnoticed.

Over the past decade, with the help of many unusually able laboratory assistants and professional colleagues, we have devised a number of methods for the study of red cell enzymes and intermediates. These methods are, in general, based on well-known principles. Indeed, in some instances, they represent merely minor modifications of published techniques. We have used the same buffer substrate and cofactor solutions whenever possible. Most of the enzyme assays have been carried out on the same hemolyzate, and in most instances the final assay volume is 1.00 ml. This manual represents a compilation of those techniques actively used in our own laboratory. This manual does not attempt to present alternative techniques or to give "fair coverage" to all available methods. Undoubtedly, some readers may be able to find techniques more suitable for their purposes by reviewing the literature, but we believe that the methods we are presenting will be found to be reliable, reproducible, and specific. In each case the conditions of the assay have been devised so as to give results which are linear with respect to hemolyzate concentration, and, where applicable, to give good recoveries. Methods for red cell enzyme assay are presented in which the concentration of substrate has been reduced so that genetic variants which appear to have normal activity at saturating substrate concentrations but which have an abnormally low affinity for substrate will be detected.

Sufficient details regarding general methodology are given in Part II so that the laboratory worker who does not have experience in biochemistry should be able to perform the procedures described in Parts III to VI.

No effort has been made to include methods for the study of all red cell constituents. Rather, coverage has been limited to those techniques with which we have had sufficient experience to recommend without qualification.

2

red cell metabolism

The chief function of the red cell is to carry oxygen from the lungs to the tissues and carbon dioxide from the tissues back to the lungs. This mission is performed in a remarkably efficient manner by hemoglobin when this heme protein is transported in a flexible disc, the erythrocyte. Hemoglobin must be maintained in the reduced state, and its environment must contain the

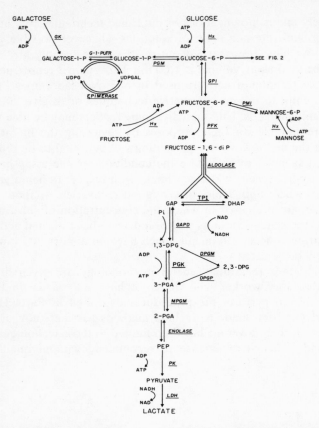

Fig. 1. The metabolism of glucose, galactose, mannose, and fructose by erythrocytes. Enzymes catalyzing the various metabolic steps are underlined. The following abbreviations are employed: GK = galactokinase; G-1-PUTR = galactose-1-P uridyl transferase; UDPG = uridine diphosphoglucose; UDPGAL = uridine diphosphogalactose; PGM = phosphoglucomutase; Hx = hexokinase; GPI = glucose phosphate isomerase; PMI = phosphomannose isomerase; PFK = phosphofructokinase; TPI = triose phosphate isomerase; GAP = D-glyceraldehyde-3-P; DHAP = dihydroxyacetone-P; GAPD = glyceraldehyde phosphate dehydrogenase; 1,3-DPG = 1,3-diphosphoglyceric acid; DPGM = diphosphoglyceromutase; 2,3-DPG = 2,3-diphosphoglyceric acid; PGK = phosphoglycerate kinase; DPGP = diphosphoglycerophosphatase; 3-PGA = 3-phosphoglyceric acid; MPGM = monophosphoglyceromutase; 2-PGA = 2-phosphoglyceric acid; PEP = phosphoenolpyruvate; PK = pyruvate kinase; LDH = lactic dehydrogenase.

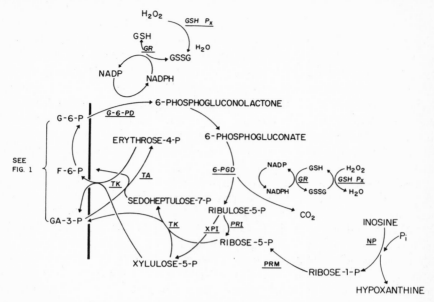

Fig. 2. The hexose monophosphate pathway of erythrocytes. Enzymes catalyzing the various pathways are underlined. The following abbreviations are used: GSH P_x = glutathione peroxidase; GSH = reduced glutathione; GR = glutathione reductase; GSSG = oxidized glutathione; G-6-P = glucose-6-P; G-6-PD = glucose-6-P dehydrogenase; F-6-P = fructose-6-P; TA = transaldolase; 6-PGD = 6-phosphogluconate dehydrogenase; TK = transketolase; GA-3-P = glyceraldehyde-3-P; PRI = phosphoribose isomerase; XPE = xylulose phosphate epimerase; PRM = phosphoribomutase; NP = nucleoside phosphorylase.

proper concentrations of various organic and inorganic ions. The metabolism of the red cell is geared to this task, and plasma glucose is its chief energy source. Anaerobic breakdown of glucose to pyruvic or lactic acid in the Embden-Meyerhoff pathway (Fig. 1) can result in the phosphorylation of ADP* to ATP and in the reduction of NAD to NADH. The primary function of ATP appears to be to power the pump which extrudes sodium and pumps potassium into the cell. It is also apparently important in maintaining the biconcave shape of the erythrocyte which is characteristic of many species, and it is required for the synthesis of certain vital red cell compounds such as FAD, AMP, and GSH. NADH is required to reduce the oxidized form of hemoglobin, methemoglobin, to reduced hemoglobin. The production of 2,3-diphosphoglycerate is another important function of the Embden-Meyerhoff pathway. Together with ATP, this phosphate ester is a regulator of the

* Abbreviations are defined in the Index.

oxygen affinity of hemoglobin. After formation of 1,3-diphosphoglyceric acid (1,3-DPG) the Embden-Meyerhoff pathway branches. 1,3-DPG can be converted to 3-phosphoglyceric acid directly through the phosphoglycerate kinase step, resulting in the phosphorylation of ADP to ATP. Alternatively, it can be converted through the action of diphosphoglycerate mutase into 2,3-DPG, which, in turn, can be catabolized to 3-phosphoglyceric acid through the action of a phosphatase. This arrangement provides a means by which a red cell can metabolize glucose with or without net gain of ATP. Thus, the ADP-phosphorylating functions of glycolysis can be carried out independently of the methemoglobin-reducing functions.

An additional pathway of glucose metabolism also exists. This is the direct oxidative shunt (hexose monophosphate shunt) (Fig. 2). In this pathway the coenzyme reduced is not NAD, as in the Embden-Meyerhoff pathway, but rather NADP. The principal function of reduced NADP (NADPH) appears to be the maintenance of glutathione in the reduced state through the glutathione reductase reaction. It can also reduce mixed disulfides of hemoglobin and GSH. Although, in vitro, NADH can also serve as a hydrogen donor, this reduced coenzyme seems quite ineffective in vivo. Glutathione appears to function in the red cells to maintain sulfhydryl groups in the active, reduced, state and through glutathione peroxidase to detoxify low levels of hydrogen peroxide. Although NADPH can also serve as a hydrogen donor for the reduction of methemoglobin, it can do so only in the presence of a dye such as methylene blue or Nile blue.

Although glucose is the principal source of energy for the red cell under normal circumstances, other substrates may also be utilized. These include inosine, fructose, mannose, and galactose. The routes of utilization of these substrates are shown in Fig. 1 (fructose, mannose, galactose) and Fig. 2 (inosine).

While the red cell does not have the capacity to produce proteins, it is able to synthesize glutathione and the coenzymes NAD, NADP, and FAD.

part II

basic techniques
and equipment

3

the preparation of red cells for assay

A. ANTICOAGULANT SOLUTIONS

Although some of the estimations described in this manual can be carried out on samples of capillary blood, it is usually most convenient to obtain a venous blood sample. The most generally satisfactory anticoagulant is acid-citrate-dextrose (ACD) solution. ACD is prepared as follows:

	Formula A	Formula B
Citric acid ($C_6H_8O_7$)	7.3 g	4.4 g
Sodium citrate ($C_6H_5O_7Na_3 \cdot 2H_2O$)	22.0 g	13.2 g
Glucose	24.5 g	14.7 g
Water to make	1000 ml	1000 ml

Four milliliters of freshly drawn venous blood are added to 0.6 ml of ACD formula A or to 1 ml of ACD formula B. Blood drawn into tubes containing 1 mg of disodium ethylenediaminetetraacetic acid (Na_2EDTA) or 10 units of heparin per milliliter of blood is also quite satisfactory.

It is often not possible to carry out assays of red cell enzymes or intermediates immediately after drawing of the blood. In some instances, it is necessary to ship the sample under conditions in which refrigeration would be cumbersome. Table I summarizes data regarding the stability of various red cell enzymes and intermediates upon storage in ACD solution, EDTA, or heparin. Almost invariably red cell constituents are much more stable when the cells are kept in whole blood than if they are washed and frozen.

B. PREPARATION OF HEMOLYZATES

Some determinations of red cell constituents are carried out on the whole blood. In the case of many of the metabolic intermediates, this is done because the contribution of plasma and other formed elements is negligible, and be-

8

Table I. Stability of Red Blood Cell Enzymes and Metabolic Intermediates in Blood Stored in Different Media*

	25° C			4° C		
	ACD	EDTA	Heparin	ACD	EDTA	Heparin
Hexokinase	5+	5+	5+	20+	20+	20+
Glucose phosphate isomerase	5+	5+	5+	20+	20+	20+
Phosphofructokinase	2	1	1	20+	6	6
Aldolase	5+	5+	5+	20+	20+	20+
Triose phosphate isomerase	2	2	2	6	6	6
Glyceraldehyde-P dehydrogenase	5+	5+	5+	6	6	6
Phosphoglycerate kinase	5+	5+	5+	20+	20+	20+
Diphosphoglycerate mutase	5	2+	5+	20+	6	6
Monophosphoglyceromutase	5+	5+	5+	20+	20+	20+
Enolase	5+	5+	5+	20+	20+	20+
Pyruvate kinase	5+	5+	5+	20+	20+	20+
Lactate dehydrogenase	5+	5+	5+	20+	20+	20+
Glucose-6-P dehydrogenase	5+	5+	5+	20+	20+	20+
6-Phosphogluconic dehydrogenase	5+	5+	5+	20+	20+	20+
Glutathione reductase	5+	5+	5+	20+	20+	20+
Glutathione peroxidase	5+	5+	5+	20+	20+	20+
NADPH diaphorase	5+	5+	5+	20+	20+	20+
NADH diaphorase	5+	5+	5+	20+	20+	20+
Phosphoglucomutase	5+	5+	5+	20+	20+	20+
GOT	5+	5+	5+	20+	20+	20+
Adenylate kinase	5+	5+	5+	20+	20+	20+
Adenosine deaminase	5+	2	2	20+	20+	20+
Galactose-1-P uridyl transferase	5+	5+	5+	20+	20+	20+
Galactokinase	5+	2	2	20+	6	6
UDP glucose-4-epimerase	1	1	1	6	6	6
Cholinesterase	5+	2	2	20+	6	6
Catalase	5+	5+	5+	20+	20+	20+
GSH	5+	1	2	20+	6	6

* The numbers indicate the number of days of storage with <10% loss of activity.

Whole blood samples were stored at room temperature (25 ± 2° C) and at 4° C under sterile conditions. Assays were carried out on room-temperature-stored samples after 1, 2, and 5 days. Samples stored at 4° C were examined at 6 days and 20 days.

Phosphorylated sugar intermediates such as glucose-6-phosphate, fructose-6-phosphate, fructose diphosphate, 2,3-DPG are very unstable in freshly drawn blood, as are the nucleotide intermediates, ATP, ADP, and AMP. Determinations for these intermediates should all be carried out on blood which is immediately deproteinized in perchloric acid.

cause marked changes in some of these substances occur during the processing of blood to separate the red cells from the plasma and other formed elements. Most enzyme assays, however, are carried out on red cells which have been carefully freed of the other formed elements and have been washed free of plasma in 0.9% sodium chloride solution. Complete removal of formed elements is particularly important in the case of assays for some enzymes, such as pyruvate kinase, in which the activity of the leukocyte enzyme may be many-fold higher than that of the erythrocyte enzyme. Since leukocyte enzyme activities may be unaffected in some of the hereditary red cell enzyme deficiencies, the existence of the enzyme deficiency may be obscured if insufficient care has been taken to remove leukocytes and platelets.

The following procedure for removal of white cells and platelets is used to prepare hemolyzates for red cell enzyme assays:

1.   A mixture of equal weights of dry α cellulose and dry microcrystalline cellulose (Sigmacell type 50, Sigma Chemical Company) is mixed with 0.154M aqueous sodium chloride solution.*

2.   The barrel is removed from a 5-ml syringe made of plastic or glass, and it is fixed in a vertical position with the outlet pointing downward. A small piece of Whatman #1 filter paper is placed at the bottom of the syringe and the cellulose slurry is poured into the syringe to the 2-ml mark. The bed is washed with 5 ml of a 0.154 M NaCl solution.

3.   One milliliter of whole, anticoagulated blood collected in EDTA, ACD or heparin is allowed to flow through the column and the effluent is collected in a centrifuge tube. The blood is washed through the column with about 1 ml of 0.154 M NaCl.

4.   The cell suspension is diluted to about 10 ml with ice-cold 0.154 M sodium chloride solution, and is centrifuged in the cold at approximately 1000 g for 15 min. Supernatant is removed from the packed red cells, taking care not to disturb the upper layer of erythrocytes and the cells are again suspended in about 10 ml of cold 0.154 M NaCl solution. The suspension is centrifuged at approximately 1000 g for 10 min and the supernatant is again removed. The washing in cold sodium chloride solution is repeated one more time, and the packed erythrocytes are suspended in approximately one volume of saline.

---

* If cellulose is not available, purified cotton, USP, can be used. It is important that USP cotton is employed, since many "cotton" preparations are actually made from rayon or other synthetic fibers. A simple test for composition is to ignite the "cotton". Genuine cotton burns leaving scarcely any ash. One thousand milligrams of cotton are boiled for 5 min with each of five changes of distilled water. It is then equilibrated with 0.9% sodium chloride solution. The cotton is placed into a column with a 1- to 1.5-cm diameter. A 1-ml blood sample is passed through the column and washed through with 1 ml of 0.9% sodium chloride. Then proceed with step 4 above.

5. One part of the suspension (usually 0.2 ml) is added to 1.8 ml of β-mercaptoethanol EDTA stabilizing solution (prepared by bringing 0.05 ml of β-mercaptoethanol and 10 ml of neutralized 10% [0.27 M] EDTA to a volume of 1 liter with water).

6. The tube containing the hemolyzate is immersed in a dry ice-acetone or dry ice-alcohol mixture until it is completely frozen and is then thawed by placing the tube into a beaker containing water at room temperature. When the hemolyzate is completely thawed, it is mixed and the tube is transferred into ice water and is maintained at 0° C.

Hemoglobin estimations are made by transferring 0.2 ml of each hemolyzate to 10 ml of ferricyanide-cyanide reagent and reading the optical density at 540 nm (see pages 11–12). While we prefer to express the results of enzyme assays in terms of activity per grams of Hb (see page 30), some investigators prefer to relate enzyme activity to the number of red cells. If this is to be done, the 50% red cell suspension is used for performance of red cell counts.

The hemolyzate prepared in this way is referred to as the 1:20 hemolyzate. It is suitable for assay of almost all of the glycolytic enzyme assays, and may be used for a working day provided that it is kept on ice. The least stable enzyme activities in this hemolyzate are glucose-6-phosphate dehydrogenase and phosphofructokinase. These assays should be conducted within 1 or 2 hours of the time that the hemolyzate is prepared.

HEMOGLOBIN ESTIMATION

Hemoglobin concentrations are measured by adding blood or hemolyzate to ferricyanide-cyanide reagent to convert the pigment to cyanmethemoglobin and then reading the optical density at 540 nm. A number of different formulations of ferricyanide-cyanide (Drabkin's solution) are in use. We find a solution containing 100 mg of NaCN and 300 mg of $K_3Fe(CN)_6$ per liter to be very satisfactory, giving stable readings within a very few minutes. Rapid readings may also be made with a formulation [1] which contains 200 mg of $K_3Fe(CN)_6$, 50 mg of KCN, 1.0 ml of 1 M KH_2PO_4, and 0.5 ml of Sterox SE (Hartman-Teddon Company, Philadelphia) per liter. A more commonly used reagent contains 200 mg of $K_3Fe(CN)_6$, 50 mg of KCN, and 1 g of $NaHCO_3$ per liter but it is necessary to wait approximately 15 min before making readings. These mixtures are stable for several months in dark bottles at room temperature. Each colorimeter or spectrophotometer used for hemoglobin estimation should be individually calibrated using a commercially available cyanmethemoglobin standard. Such a standard should be diluted in ferricyanide-cyanide solution to provide a range of concentrations from approximately 5 mg % to approximately 45 mg % (0.005 to 0.045 g %). It is important to bear in mind that these are the actual concentrations in the cuvette. (Some standards are labeled by the manufacturer with concentrations which are 250-fold or 500-fold higher than the actual concentration of hemoglobin in the standard in order to represent the hemoglobin concentrations of a blood sample before dilution in ferricyanide-cyanide reagent.) The optical density of the standards is read against a ferricyanide-cyanide blank. When the absorbance of the diluted standard solutions is plotted against their hemoglobin concentrations, a straight line should be obtained. For ease of calculation, the absorbance value (OD_{540}) at which the calibration curve intersects the hemoglobin concentration of 10 mg % is designated A_1. The hemoglobin calibration factor (F_{HB}) is then

$$\frac{1}{100A_1}$$

The concentration of hemoglobin in the cuvette in grams per 100 ml is

$$Hb = F_{HB} \times OD_{540}$$

The concentration of hemoglobin in grams per 100 ml in any sample is

$$Hb = \frac{OD_{540} \times F_{HB} \times (V_{Fe} + V_{Hb})}{V_{Hb}}$$

where V_{Fe} is the volume of ferricyanide-cyanide solution used and V_{Hb} is the volume of sample added to the ferricyanide-cyanide solution.

The concentration of hemoglobin in grams per liter (10 Hb)

$$10\ Hb = \frac{OD_{540} \times F_{HB} \times (V_{Fe} + V_{Hb}) \times 10}{V_{Hb}}$$

and in grams per milliliter (0.01 Hb)

$$0.01\ Hb = \frac{OD_{540} \times F_{HB} \times (V_{Fe} + V_{Hb})}{100 V_{Hb}}$$

D. RED CELL EXTRACTS

The preparation of extracts of red cells is convenient in the determination of many metabolic intermediates. In general, one of five different extraction methods may be employed: (1) boiled extracts, (2) trichloroacetic acid extracts, (3) perchloric acid extracts, (4) $Ba(OH)_2$-$ZnSO_4$ extracts, and (5) metaphosphoric acid extracts. Each has its own advantages and disadvantages.

In calculating the concentration of a substance in the red cell, it is necessary to make appropriate corrections for the degree of dilution during the extraction process. When the hemolyzate is very dilute, this is a relatively simple matter. When a concentrated preparation of red cells is treated with a small volume of precipitating agent, however, a substantial correction must be made for the fact that a portion of the red blood cells is a solid, mostly protein. Thus, after precipitation of proteins, the water-soluble substances are diluted into a fluid volume which is substantially smaller than the sum of the volume that has been treated and the volume of the precipitating solution. It is convenient, and relatively accurate, to assume that each milliliter of packed red cells contains 0.7 ml of water. Whole blood is assumed to contain 0.8 ml of water per milliliter.

The dilution (D) of a substance during the precipitation of packed red cells is then obtained by dividing the volume of the red cells (V_R) by the sum of the volume of the precipitating solution (V_P), the volume of the water or saline in which the cells are suspended (V_W), and the red cell water, viz., 0.7 times the volume of the red cells:

$$D = \frac{V_R}{V_P + V_W + 0.7 V_R}$$

The denominator of this expression represents the total fluid volume into which the substance has been diluted during the extraction process. Since the numerator represents the volume of packed cells, dividing the concentration of a substance in the extract by the dilution factor will give a concentration which represents the concentration of material in the packed red cells, not in the red cell water.

The concentration of a substance in red cell water may be approximated by dividing the concentration in red cells by 0.7, assuming that the distribution of the substance is limited to the water phase. If an extract of whole blood is made, calculation of the dilution factor is carried out as follows:

$$D = \frac{V_B}{V_P + 0.8 V_B}$$

where V_B is the volume of blood. Here, dividing the concentration of a substance in the extract by D will give its concentration in the blood sample. If the substance is limited to the red cells, the concentration in red cells would be estimated by dividing the concentration in whole blood by the hematocrit. The calculation of dilution during extraction is illustrated by the following four examples.

Example 1. Five milliliters of a 50% red cell suspension are treated with 1 ml of a precipitating agent. The water in this system is comprised of 1.0 ml (precipitating agent) + 2.5 ml (aqueous phase of 50% cell suspension) + 0.7 × 2.5 ml (aqueous phase of red cells) = 5.25 ml. The dilution of red cells is not 2.5 to 6, as might have been thought, but rather 2.5 to 5.25:

$$D = \frac{2.5}{1 + 2.5 + (0.7 \times 2.5)} = 0.476$$

The dilution factor is therefore 0.476. The concentration of a metabolite in the extract was found to be 2.6 μM. The concentration in red cells is therefore 5.46 μM and the concentration in red cell water is 5.46 ÷ 0.7 or 7.8 μM.

Example 2. Five milliliters of whole blood are treated with 3 ml of a precipitating reagent:

$$D = \frac{5}{3 + (0.8 \times 5)} = 0.714$$

The pyruvate level in the extract, after correction for dilution during neutralization and assay, was found to be 123 μM. The blood pyruvate level is therefore 172.2 μM.

Example 3. Five milliliters of a 1:100 dilution of red cells are precipitated with 5 ml of precipitating solution. In this case, the red cell sample is so dilute that red cell solids may be ignored. The dilution is therefore 1:200. If one wished to make the exact calculation, the correct dilution would be 0.05:9.985, or 1:199.7.

Example 4. Five milliliters of a hemolyzate prepared from a 50% red cell suspension are extracted with 5 ml of a precipitating reagent. The supernatant solution is removed, and 10 ml of half-strength precipitating reagent are added to the precipitate, mixed, and removed after recentrifugation. The pooled filtrates occupy a volume of 16.2 ml. In this case one may assume that virtually all of the constituent to be measured has been removed from the red cells, and the dilution is therefore 2.5:16.2. However, if very precise results are desired, an additional extraction of the precipitate would be desirable. This may dilute the constituent to be measured to too great an extent; alternatively, one might wash the precipitate twice with 5-ml aliquots, resulting in more complete extraction.

1. Boiled extracts

Boiling red cells is the easiest way to produce a protein-poor extract. No foreign substances are added, and the technique is especially suitable when acid-labile components are being studied. One of the chief disadvantages of boiled extracts is that their optical clarity is inferior to that of other extracts. If the concentration of the material to be used is very high or if the analytic method is very sensitive, so that the procedure can be carried out at a dilution of 1:10 or greater in the cuvette, this poses no serious problem.

Usually a Pyrex or heat-resistant plastic centrifuge tube containing diluted hemolyzed red cells is placed in a boiling water bath and stirred vigorously with a glass stirring rod. It is advantageous to adjust the NaCl concentration of the sample to 0.15 M. This results in the formation of a coarser precipitate, which is more easily removed by centrifugation. In the case of dilute hemolyzates (1:10 or greater), 1 min of boiling is usually sufficient. When very con-

Table II. Minimum Concentration of Precipitating Agents Required to Produce a Satisfactory Extract

Concentration of blood (%)	TCA: blood = 1:2	TCA: blood = 1:1	TCA: blood = 4:1	Concentration of packed red cells (%)	TCA: cells = 1:2	TCA: cells = 1:1	TCA: cells = 4:1
100	20%	10%	5%	100	20%	10%	5%
50	20%	10%	2.5%	50	20%	10%	5%
25	10%	5%	2.5%	25	10%	5%	2.5%
10	10%	5%	2.5%	10	10%	5%	2.5%
1	10%	5%	2.5%	1	10%	5%	2.5%
	PCA: blood = 1:2	PCA: blood = 1:1	PCA: blood = 4:1		PCA: cells = 1:2	PCA: cells = 1:1	PCA: cells = 4:1
100	8%	4%	2%	100	8%	4%	2%
50	4%	2%	1%	50	4%	4%	1%
25	4%	2%	1%	25	4%	4%	1%
10	4%	2%	1%	10	2%	2%	1%
1	4%	2%	1%	1	2%	2%	1%

Normal washed packed red cells or whole blood with hematocrit adjusted to 50% was used.

The following concentrations (W/V) of TCA were studied: 50%, 20%, 10%, 5%, 2.5%.

The following concentrations of PCA were studied: 20%, 10%, 8%, 6%, 4%, 2%, 1%, 0.5%.

The lowest concentration producing a satisfactory filtrate is given in the table. In practice, it is advisable to use a somewhat higher concentration so that one may be certain to obtain a clear filtrate from every sample.

centrated red cells are boiled, complete heat denaturation may require considerably longer exposure to heat since the coagulum forming at the hot sides of the tube tends to insulate the central portion of the sample. After heating, the sample is rapidly cooled in ice, centrifuged, and the supernatant fluid is removed.

2. Trichloroacetic acid (TCA) extracts

Trichloroacetic acid is a suitable agent when extracts of good optical clarity are desired, and when these extracts are to be studied using enzymatic techniques. The concentration of TCA required depends on whether blood or red cells are being extracted, on the dilution of the samples, and on the ratio of TCA to hemolyzate. The minimum concentrations required for a normal sample are given in Table II. The amount may vary, of course, depending on the hematocrit of a blood sample, the serum lipid concentration, and the age of the TCA solution. For most purposes, the TCA and hemolyzate should both be ice cold and the TCA solution added while the hemolyzate is being stirred constantly. After standing for a few minutes, the mixture is centrifuged in the cold at 1000 to 2000 g and the supernatant fluid removed. It may, in some instances, be useful to add one or more aliquots of TCA solution to the precipitate, mix well, recentrifuge, and pool the supernatants with the first extract. The TCA may now be removed from the extract by treatment with ether. This is conveniently done by adding several volumes of ether to the TCA extract, capping the tube with a cork, glass, or rubber stopper (not Parafilm), and then shaking vigorously or mixing on a Vortex mixer (e.g., Scientific Products S8220). After standing for a few minutes in the cold, the ether (top) phase will separate from the aqueous (bottom) phase and the ether may be aspirated and discarded. The process should be repeated three or four times, and residual ether may then be removed with a stream of air or nitrogen in a ventilated hood. It is convenient to hold the tube in the left hand during this process, bubbling gas gently through the solution with a Pasteur pipet. At first, the solution begins to feel very cold; when it rewarms, the ether has been driven off. If removal of TCA by ether has been complete, the pH of the extract should be above 6. This may be verified with pH paper. If the pH is below 6, some residual TCA may still be present but will not necessarily interfere with subsequent manipulations. The extract may also continue to maintain a low pH if it contains substantial quantities of weak acids which are not ether soluble.

3. Perchloric acid (PCA) extracts

Perchloric acid produces a particularly clear extract. It is relatively easy to use, but calculations may be somewhat complicated because of various volume adjustments which may be required.

PCA is suitable for the extraction of somewhat acid-labile compounds, such as ATP, but not for extremely acid-labile materials, such as NADPH or NADH, or easily oxidized compounds, such as GSH. Ice-cold perchloric acid solution is added to whole blood or hemolyzate in the ratio desired (Table II). After standing for a few minutes, the mixture is centrifuged in the cold and a carefully measured volume of the supernatant fluid is transferred into a calibrated tube. As in the case of extraction with TCA, a second extraction may be carried out if desired and pooled with the first. A drop of 0.05% W/V methyl orange solution is added directly to the PCA supernatant. The ice-cold extract is now neutralized with a 1 to 3 M solution of potassium carbonate. If dilution of the sample is not a problem, the more dilute carbonate solution may be used and is often more convenient because of the smaller likelihood of overtitrating the extract. The dropwise addition of potassium carbonate will result in effervescence (as carbon dioxide evolves) and in the formation of a heavy white precipitate. The tube should be shaken vigorously between additions, so that the reaction with each drop will proceed to completion before the next drop is added. When the color of the indicator changes to yellow, adequate neutralization has been achieved. The neutralized solution is adjusted to a fixed volume with water. The precipitate will generally settle out spontaneously within a few moments, and further centrifugation is not required except for fluorometric assays. The supernatant fluid is relatively free of perchlorate and is suitable for most assays.

4. Barium hydroxide-zinc sulfate extracts

$Ba(OH)_2$-$ZnSO_4$ extracts of red cells or whole blood are particularly useful for the study of sugars, such as glucose and galactose. Solutions of approximately 0.15 M $ZnSO_4$ and 0.15 M $Ba(OH)_2$ are prepared. These reagents should be titrated against one another to assure that equal volumes neutralize each other exactly. This is done by measuring 10 ml of the 0.15 M $ZnSO_4$ into a flask containing 50 ml of water, adding 4 drops of 0.2% alcoholic phenolphthalein indicator, and slowly titrating with the 0.15 M $Ba(OH)_2$ until 1 drop turns the solution a faint pink. If less than 10 ml of 0.15 M $Ba(OH)_2$ are required, dilute the $Ba(OH)_2$ with distilled water so that

$$\frac{V_I}{V_F} = \frac{v}{10} \quad \text{or} \quad V_F = \frac{10\,V_I}{v}$$

where V_I equals the volume of undiluted $Ba(OH)_2$ solution, V_F is the volume of the solution after dilution, and v is the volume of undiluted $Ba(OH)_2$ solution required to neutralize 10 ml of $ZnSO_4$.

If more than 10 ml of $Ba(OH_2)$ solution are required, the $ZnSO_4$ solution is diluted so that

$$\frac{V_I}{V_F} = \frac{10}{v} \quad \text{or} \quad V_F = \frac{V_I\,v}{10}$$

where V_I is the volume of the undiluted $ZnSO_4$ solution, V_F is the volume of the solution after dilution, and v is the volume of $Ba(OH)_2$ solution which was used to neutralize the undiluted $ZnSO_4$ solution.

The adjusted solutions are stored in tightly stoppered bottles. The $Ba(OH)_2$ bottle should be fitted with a double-hole stopper. A calcium chloride tube reaching the air space above the solution is placed in one hole, while a glass tube reaching to the bottom of the bottle is placed in the other hole. Tightly fitting plastic tubing is attached to the glass tube and is sealed with a clamp. $Ba(OH)_2$ samples may then be withdrawn with a syringe through the plastic tube, and the loss of $Ba(OH)_2$ content due to formation of carbonic acid is prevented because air entering the bottle is filtered through the calcium chloride.

Extracts may be prepared from either blood or red cells. Pipet 1.5 ml of distilled water and 0.1 ml of blood or red cells into a centrifuge tube, and mix. Add 0.2 ml of $Ba(OH)_2$ solution, mix, and allow to stand 1 min. Add 0.2 ml of $ZnSO_4$ solution, mix, and allow to stand 3 to 5 min. Centrifuge 10 min at 1000 g, and decant the supernatant.

5. Metaphosphoric acid extracts

Metaphosphoric acid-salt-EDTA is a useful agent for the extraction of acid-stable materials which are to be determined with a nonenzymatic technique with which these materials do not interfere. It is particularly useful in the determination of red cell GSH levels because contaminating metallic ions are chelated. The precipitating solution contains 1.67 g of glacial metaphosphoric acid (a mixture of HPO_3 and $NaPO_3$), 30 g of NaCl, and 0.2 g Na_2- or K_2EDTA per 100 ml [2]. The precipitating solution is not clear because of the presence of a fine precipitate of undissolved EDTA, and is stable for approximately 3 weeks in the cold. Three volumes of precipitating solution are added to two volumes of a 1:4 or greater dilution of a whole blood and mixed thoroughly. The mixture is allowed to stand for a few minutes and filtered through coarse or medium filter paper.

REFERENCES

1. van Kampen, E. J., and Zijlstra, W. G.: Standardization of hemoglobinometry. II. The hemiglobincyanide method. Clin. Chim. Acta 6:538–545, 1961.
2. Beutler, E., Duron, O., and Kelly, B. M.: Improved method for the determination of blood glutathione. J. Lab. Clin. Med. 61:882–890, 1963.

4

reagents

A. SOURCES

In the course of our studies we have used reagents from a variety of commercial sources. These have included, but have not necessarily been limited to, the following:

1. Sigma Chemical Company
    3500 DeKalb Street
    St. Louis, Mo. 63118

2. Boehringer Mannheim
    C. F. Boehringer & Soehne GMBH, Manheim, West Germany
    582 Market Street, San Francisco, Calif. 94104
    20 Vesey Street, New York, N.Y. 10007

3. Calbiochem
    10933 N. Torrey Pines Road
    La Jolla, Calif. 92037

It is usually immaterial whether the free acid or a sodium, potassium, or other salt of organic reagents is used, but barium and other heavy metal salts must not be employed. Thus it does not matter whether 6-phosphogluconic acid or 6-phosphogluconate, L-aspartic acid or L-aspartate, etc., is used, as long as the amount used is small or the pH is adjusted prior to use.

In general, we have encountered no serious difficulties with commercial reagents, except for enzyme preparations and for ^{14}C-galactose. The enzymes which are purchased will vary somewhat from the labeled potency, especially after they have been stored for some time. More important, traces of contaminating enzymes are often present. It is common, for example, for glucose-6-P dehydrogenase to be contaminated with hexokinase, and for α-glycerophosphate dehydrogenase to contain triose phosphate isomerase. If one of the contaminants is the enzyme which is being assayed and if appropriate corrections are not made, serious errors could arise. Blank determinations, in which β-mercaptoethanol-EDTA stabilizing solution is substituted for hemolyzate, serve this purpose when auxiliary enzymes are used in an assay, and must always be carried out when a new lot of enzyme is used. If the blank rate is appreciable, it must be estimated with each assay and subtracted from the value obtained with hemolyzate.

The substrates used for enzyme assays also often contain various contaminants which may not be listed on the label. The extent of contamination of a group of substrates was measured recently [1]. Fructose-6-phosphate and 2,3-DPG are particularly likely to be impure, but the possibility that contaminants may be present in any reagent which has been purchased must always be considered, particularly when anomalous results are encountered.

In those instances in which concentration or purity of reagents is critical and the purity of the commercial reagents is insufficient, we standardize the concentration of the reagents prior to their use or carry out special purification procedures. The methods employed are described in Parts III, IV, and V.

B. PREPARATION

With rare exceptions, quantities are expressed as concentration in M (moles/liter), mM (millimoles/liter; mmoles/liter), or μM (micromoles/liter; μmoles/liter). Since there is seldom an occasion to prepare reagents for enzyme assays in liter quantities, it is useful to remember that moles/liter (M) equals mmoles/ml, and that mmoles/liter (mM) equals μmoles/ml. Unless otherwise indicated, all solutions used in this manual are prepared in water. Enzymes are usually diluted in β-mercaptoethanol-EDTA stabilizing solution.

In order to prepare a solution of a given molarity, the molecular weight is multiplied by the molarity of the solution desired to give the number of grams/liter (or milligrams/milliliter) needed. Often the molecular weight of a reagent is given on the container or in the catalogue from which it was ordered. Sometimes other sources must be consulted. The *Merck Index* [2] is particularly useful. Sometimes it is necessary to calculate the molecular weight from the structural formula, adding up the atomic weights of each of the atoms in the compound. A list of frequently used molecular weights is given in Appendix 4. If the molecular weight is calculated or found in a reference source, it is important to be certain that the chemical form listed is the same one which has been purchased. If, for example, the molecular weight of the free acid of a compound is 323, the molecular weight of the disodium salt would be 367: $323 + (2 \times 23)$ (the atomic weight of sodium) $- (2 \times 1)$ (the atomic weight of hydrogen). Similarly, it is necessary to adjust the molecular weight for water of hydration. Finally, if the purity of the substance used is known, the molecular weight should be adjusted for purity prior to preparation of solutions. This is accomplished by dividing the molecular weight by the purity, 1.0 being 100% pure, 0.9 being 90% pure, etc. When the quantities of reagents to be prepared are small, as is usually the case in the procedures described in this manual, it is a waste of time to attempt to weigh exact, small quantities of reagents to bring to a fixed volume. Rather, the concentration per milliliter (c) desired should be multiplied by the number of milliliters of reagent required and an exact weighing be made of approximately this quantity. The volume of water to be added (v) is then simply obtained by dividing the exact quantity which has been weighted (w) by the concentration per milliliter (c) desired: $v = w/c$. The effect of the substance itself on the total volume can be ignored unless a very concentrated solution is being prepared, in which case it must be brought to the desired volume in calibrated glassware.

Example. You wish to prepare a 2 mM solution of NADP, and find that the anhydrous weight of NADP is 743. But you have purchased the monosodium salt containing four molecules of water, and the label states that the preparation is 94% pure. The molecular weight of monosodium NADP with four molecules of water is, then, $743 + 23 - 1 + (4 \times 18) = 837$. The molecular weight adjusted for purity is

837/0.94 = 890. A 2 mM solution should then contain 2 × 890 μg/ml, or 1.78 mg/ml. You have decided that you want to make enough NADP solution for about 40 assays. This will require about 8 ml of the solution, and so you attempt to weigh out about 14 mg. The actual amount weighed turns out to be 16.93 mg. This is a sufficient quantity for 16.93/1.78 ml. Accordingly, you pipet 9.51 ml of distilled, deionized water into the tube containing the NADP. Alternatively, a calibrated 10-ml cylinder would be sufficiently accurate.

NADPH and NADH are unstable, not only in solution but even in the dry state, at $-20°$ C. When the concentration of the reduced pyridine nucleotides is critical, a solution more concentrated than the desired final concentration should be prepared and assayed spectrophotometrically. This is required, in the procedures described in this manual, only in the assay for glutathione reductase. A NADH or NADPH solution containing approximately 2 mg/ml is prepared. Eighty-five hundredths milliliter of water and 0.1 ml of 1 M tris-HCl-EDTA buffer, pH 8, are added to a cuvette, and its optical density (A_0) is measured at 340 nm against air or against a water blank. Five hundredths milliliter of the NADH or NADPH solution is added, and a second reading (A_1) is taken. The concentration of the pyridine nucleotide in the solution is then

$$C = \frac{A_1 - A_0}{0.311}$$

where C is the concentration (mM). To dilute the solution to give a 2 mM concentration, 2 ml are brought to a total volume of C ml.

Commercial suppliers of enzymes often label their preparation in terms of the number of micromoles of substrate converted per minute under stated conditions of assay. One unit of enzyme is that quantity of enzyme which converts 1 μmole of substrate per minute. Enzymes are usually, although not always, supplied as ammonium sulfate suspensions. They should usually be stored in fully concentrated form and only diluted on the day of use. One exception to this rule is the preparation of a partially diluted enzyme suspension for the purpose of carrying out phosphofructokinase and aldolase assays. It should be noted, however, that even here only the ammonium sulfate-stabilized, mixed suspension is stored. Aqueous UDPG dehydrogenase is also fairly stable. Usually, however, dilutions of enzymes are discarded at the end of a working day. Auxiliary enzymes, should be diluted in the β-mercaptoethanol-EDTA stabilizing solution and be kept on ice.

C. BUFFERS

In preparing buffers it is important to bear in mind that the pH of a buffer is influenced by both changes in temperature and by dilution. The pH of buffers given in this manual is that at the initial concentration of the buffer at 25° C.

Tris buffers are prepared by weighing the amount of tris needed for the volume of buffer desired at the specified molarity and bringing to approximately 80% of the final volume with water, at room temperature. While the tris solution is being stirred continually with a magnetic stirrer, its pH is monitored with a pH meter and acid is added until the desired pH value has been achieved. The buffer is then brought to the exact volume desired. The concentration of acid is immaterial with regard to the final composition. It must be sufficiently strong so that the final volume desired is not exceeded, but sufficiently weak so that abrupt pH changes do not occur. It is often useful to adjust the pH to within 0.2 or 0.3 units of the desired value with concentrated acid, and then to make the final, fine adjustment with a much more dilute solution. For the preparation of tris-hydrochloride buffers, hydrochloric acid is used; for tris-acetate buffers, acetic acid is used, and so on.

Citrate buffers are prepared by weighing out the needed amount of citric acid, proceeding as for tris buffers but adjusting the pH with a NaOH solution of convenient concentration.

Example. To prepare 100 ml of 1 M tris-HCl buffer, pH 8.0, with 5 mM EDTA 12.1 g of tris and 168 mg of disodium EDTA are placed in a beaker and dissolved in approximately 80 ml of distilled water at room temperature. Concentrated HCl is added until the pH as measured with a pH meter is about 8.2. The pH is now adjusted to exactly 8.0 using 2 M HCl. The mixture is then carefully transferred to a 100-ml volumetric flask. The empty beaker is rinsed with a little distilled water, which is poured into the volumetric flask, and the final volume is adjusted to 100 ml.

Phosphate buffers are prepared from stock solutions of 1 M KH_2PO_4 and 1 M K_2HPO_4. A buffer table (Table III) is used.

Example. One hundred milliliters of a 0.1 M phosphate buffer, pH 7.4, are required. Entering the table at 0.1 M, we find that 0.802 part of K_2HPO_4 (and therefore 0.198 part of KH_2PO_4) is required. Since the final molarity is to be 0.1 M, 8.02 ml of the 1 M stock solution of K_2HPO_4 and 1.98 ml of the 1 M stock solution of KH_2PO_4 are pipetted into a volumetric flask and the volume brought to 100 ml. It is worth keeping in mind the fact that, if this buffer is diluted 1:10 in an assay system, the pH will be between 7.6 and 7.7 (see Table III).

D. WATER

Many enzymes are inhibited by trace substances which are commonly found in water. Erratic laboratory results can sometimes be traced to impurities in the water used to prepare reagents. Metal ions which have gained access to the water supply, either through stills or through piping, can be particularly troublesome. We have obtained very satisfactory results by distilling ordinary tap water in a metal still, and then passing this water through a D 5040 (Scientific Products) deionizing cartridge, permitting the resistance to fall no

Table III. The Preparation of Phosphate Buffers of Different pH Values and Molar Strengths [3]

pH	Total concentration of phosphate (moles/liter)										
	0	0.01	0.04	0.10	0.20	0.30	0.40	0.50	0.60	0.80	1.0
	Mole fraction of total phosphate as K_2HPO_4										
5.5									.099	.119	.134
5.6								.109	.121	.141	.157
5.7						.104	.121	.132	.145	.165	.182
5.8				.085	.110	.129	.146	.158	.171	.192	.212
5.9	.048	.065	.083	.106	.135	.155	.173	.186	.200	.224	.244
6.0	.059	.081	.103	.132	.163	.185	.203	.219	.236	.259	.277
6.1	.074	.100	.126	.160	.195	.220	.239	.256	.273	.295	.312
6.2	.091	.122	.155	.192	.232	.261	.281	.298	.312	.333	.349
6.3	.112	.150	.190	.232	.276	.305	.326	.341	.354	.372	.386
6.4	.137	.183	.230	.278	.325	.353	.373	.385	.398	.414	.424
6.5	.166	.222	.274	.328	.376	.403	.421	.435	.444	.458	.466
6.6	.201	.266	.325	.381	.429	.457	.473	.484	.493	.503	.508
6.7	.240	.315	.380	.438	.486	.511	.526	.535	.543	.549	.551
6.8	.285	.369	.440	.497	.543	.565	.578	.586	.590	.594	.594
6.9	.334	.425	.498	.557	.598	.617	.629	.634	.637	.638	.636
7.0	.387	.484	.556	.615	.651	.669	.677	.681	.683	.681	.676
7.1	.443	.544	.614	.668	.701	.716	.722	.724	.725	.721	.715
7.2	.500	.604	.670	.717	.747	.758	.764	.763	.762	.758	.751
7.3	.557	.659	.720	.762	.785	.796	.801	.800	.797	.790	.784
7.4	.613	.710	.763	.802	.822	.829	.832	.832	.828	.821	.814
7.5	.666	.756	.805	.837	.854	.860	.860	.859	.855	.848	.840
7.6	.715	.796	.840	.866	.880	.883	.884	.883	.879	.872	.864
7.7	.759	.831	.869	.890	.902	.905	.905	.904	.901	.894	.885

lower than 10^6 ohms. With increasing recognition of the importance of pure water in carrying out enzyme assays, a number of commercially available systems for purifying and deionizing water have become available. Very likely these, too, are quite satisfactory, but we have had no personal experience with them.

E. GLASSWARE

All laboratory glassware should be carefully cleaned. We pass all glassware through dichromic acid and rinse copiously, first with tap water, then with distilled water, and finally with distilled, deionized water. Cuvette washers of the type illustrated in Fig. 3 are particularly useful when small cuvettes are employed. If cuvettes are cleaned promptly after each assay, it is usually

Fig. 3. Two cuvettes with a critical volume of less than 1 ml being washed on a cuvette washer.

sufficient to rinse them with distilled, deionized water, then acetone, and then to permit them to dry. Sometimes it is necessary to use a detergent solution, and occasionally the cuvette should be permitted to stand overnight with dichromic acid.

F. STORAGE OF REAGENTS (TABLE IV)

With certain exceptions, to be discussed below, it is convenient to store all reagents in the frozen state. Although some reagents, for example $MgCl_2$ and EDTA, are quite stable even at room temperature, freezing prevents bacterial growth and mold formation. It is very important to be certain that reagents are completely thawed and then well mixed before use, because layering of water and solute occurs, resulting in variations in the concentration of the reagent removed from the bottle.

There are some reagents which should always be stored at 4° C. NAD and NADP are more stable at 4° than at —20° [4]. Most enzyme solutions or suspensions are more stable at 4° C than at —20° C, unless the supplier's instructions specify frozen storage.

Table IV. Storage Stability of Reagents

Reagents stable for several months at $-20°$ C (frozen)

Tris buffers
Phosphate buffers
Glycine buffers
Nicotinamide-glycylglycine buffers
$MgCl_2$
EDTA
ATP
ADP
Galactose
Galactose-1-P
Glucose
Glucose-6-P
6-Phosphogluconic acid
Fructose-6-P
Fructose-1,6-diP
Glucose-1,6-diP
Glucose-1-P
DL-Glyceraldehyde-3-P
2-Phosphoglyceric acid
3-Phosphoglyceric acid
2,3-Diphosphoglycerate
Phosphoenolpyruvate

UDPG
Sodium pyruvate
Pyridoxal 5'-phosphate
ACD solution
Citrate
GSSG
Disodium arsenate
KCl
KH_2PO_4 and K_2HPO_4
$K_3Fe(CN)_6$
L-Aspartate
Monosodium α-ketoglutaric acid
Methylene blue
Sodium azide
UDPG dehydrogenase (several weeks)
Saponin
NaF
^{14}C-galactose
UDPG
Malic dehydrogenase, desiccated
Ethanol
Ferricyanide-cyanide reagent
Acetylthiocholine

Reagents stable for several months at $4°$ C (if no mold forms)

Tris buffers
Phosphate buffers
Glycine buffers

$MgCl_2$
EDTA
NAD^+
$NADP^+$
Glucose
Glucose-6-P
Galactose
ACD solution
Citrate
Disodium arsenate
2,3-Diphosphoglycerate
Phosphoenolpyruvate

Sodium pyruvate
GSSG
L-Aspartate
KCl
KH_2PO_4 and K_2HPO_4
$K_3Fe(CN)_6$
Sodium azide
Methylene blue
Bovine serum albumin
30% H_2O_2
UDPG
Saponin
Malic dehydrogenase, in $(NH_4)_2SO_4$
Ethanol
NaF
Ferricyanide-cyanide reagent
DTNB

Reagents stable for 2 weeks at $4°$ C

Hb substrate for NADH diaphorase
β-Mercaptoethanol-EDTA stabilizing solution

Reagents to be prepared daily

NADH
NADPH
FAD

GSH
t-Butyl hydroperoxide
Dilute H_2O_2

Finally, there are some types of solutions which should not be stored at all. These include solutions of NADH and NADPH, diluted auxiliary enzyme solutions, and solutions of GSH and flavin adenine dinucleotide (FAD). Sulfhydryl reagents (such as solutions containing β-mercaptoethanol) should be kept for only 2 to 3 weeks in the refrigerated state.

REFERENCES

1. Niessner, H., and Beutler, E.: Contamination of commercially available intermediates of the glycolytic pathway. Experientia 29:268–270, 1973.
2. The Merck Index, Rahway, N.J., Merck & Co., Inc.
3. Green, A. A.: The preparation of acetate and phosphate buffer solutions of known pH and ionic strength. J. Amer. Chem. Soc. 55:2331–2336, 1933.
4. Lowry, O. H., Passonneau, J. V., and Rock, M. K.: The stability of pyridine nucleotides. J. Biol. Chem. 236:2756–2759, 1961.

5

instrumentation

Three basic types of measuring instruments are employed to determine the activity of most of the enzymes and concentration of various substrates in the red cell. These are an ultraviolet spectrophotometer, a colorimeter, and a filter fluorometer. Isotope counting equipment is needed to assay galactokinase activity.

Most of the techniques described are spectrophotometric and depend on the light absorption of reduced pyridine nucleotides (NADPH or NADH) at 340 nm. In estimating 2,3-DPG the concentration of phosphoenolpyruvate (PEP) is measured by its absorption at 240 nm. In fact, the absorption of PEP at 200 nm is considerably greater, but conventionally the 240-nm wavelength is used because of the better stability of most spectrophotometers at somewhat higher wavelengths.

When enzymes with relatively high activity such as glucose-6-phosphate dehydrogenase, triose phosphate isomerase, or phosphoglycerate kinase are measured, relatively large changes in optical density are observed when only very small amounts of hemolyzate are present in the cuvette. Under these conditions very highly sensitive detecting instruments are not required. Adequate assays can be obtained with the use of relatively insensitive spectrophotometers such as the Beckman DU, Beckman DB, Beckman Model B, or some of the filter photometers produced by Eppendorf. When relatively inactive enzymes are assayed, such as hexokinase, aldolase, diphosphoglycerate mutase, or epimerase, small changes of optical density are measured against a high baseline absorbance produced by the large amount of hemolyzate which must be added to the cuvette. Under these circumstances very sensitive measuring devices are required. All of our studies have been carried out with the Gilford Model 2000 or Gilford Model 2400 spectrophotometer, but there are other instruments which would undoubtedly also be highly satisfactory. In any case, it is very desirable to employ a recording spectrophotometer, so that deviations from linearity can easily be appreciated and the appropriate slope of the curve can be measured. A device which automatically adjusts the blank reading to zero, as is found in the Gilford automatic recording spectrophotometer or in double-beam spectrophotometers, is also a great convenience. The cuvette compartment should be temperature controlled. This is achieved by circulating water from a controlled-temperature water bath through plates which surround the cuvette compartment. For the most accurate results, it is desirable to measure the actual temperature in cuvettes since there will be some heat loss while the water is passing through tubing from the temperature regulator to the cuvette compartment. The temperature in the cuvette compartment will therefore be slightly lower than the temperature of the water bath when assays are carried out at $37°$. It will usually be found to be necessary to adjust the temperature of the water bath to about $39°$ C in order to achieve a $37°$ reading in the cuvette compartment, but the exact temperature required will depend on the length and type of tubing used, and on the rate of circulation.

Most of the spectrophotometric procedures depend on observing an alteration in the optical density of the contents of a single cuvette. The use of matched cuvettes is therefore not required. Glass cuvettes are entirely satisfactory for readings at 340 nm, but quartz or silica glass must be used for readings carried out at 240 nm. Any convenient size of cuvette can be used for spectrophotometric assays. We prefer, however, to use cuvettes with a critical volume of less than 1.0 ml, and we prefer to use assay systems containing a final volume of 1.00 ml. The Pyrocell cuvette 1007 (S22-260) (Pyrocell Manufacturing Company, 91 Carver Avenue, Westwood, N.J.) and the Rho Scientific SM 69 (Rho Scientific, Inc., P.O. Box 295, Commack, N.Y.) are such cuvettes, and we use them for most of our spectrophotometric assays.

The use of such cuvettes has two major advantages. First of all, it permits the use of smaller quantities of expensive reagents than does the more commonly used 10×10-mm square cuvette, but is still easy to clean when compared to cuvettes designed for use with even smaller volumes. Second, the use of a 1.0-ml volume simplifies computations. On the other hand, such cuvettes are more expensive than some of the 3.0-ml type.

The critical volume of a cuvette will vary with the cuvette carrier and instrument employed. To determine the critical volume of a cuvette in the system employed, prepare a dilution of approximately 1:200 of whole blood in ferricyanide-cyanide reagent. Place in the cuvette a measured volume which is known to be less than the critical volume of the cuvette. Read the optical density of the system at 540 nm against air, and add a small, measured increment of the solution. Read the optical density again, and repeat the procedure until further addition of the solution produces no further change in the optical density. The smallest total volume which produces the final reading is the critical volume of the cuvette. This volume or greater volume must be present in all assayed mixtures used in the spectrophotometric system. One should be certain that the aperture used for collimating the light beam by the spectrophotometer is sufficiently narrow so that it will not strike the thick sides of the cuvette. This can be done by darkening the room, opening the cuvette compartment, and observing the passage of a 500- to 600-nm light beam through a cuvette containing water.

In assays carried out at 340 nm a tungsten lamp is entirely satisfactory; at 240 nm a hydrogen or deuterium lamp must be used. It is important that the spectrophotometer be properly calibrated. In the Gilford Model 2000 or 2400, this is accomplished by using commercially available filters. The calibration of the recorder must be properly adjusted also.

In properly adjusted spectrophotometers a 0.1 mM solution (0.1 μmole/ml) of NADH or NADPH has an optical density of 0.622 at 340 nm. The slit width of the instrument is not of critical importance, since the absorbance band is fairly wide. However, at slit widths larger than 1 mm a slightly lower extinction coefficient is observed so that at 2 mm (half intensity bandwidth 13.5 nm) a 0.1 mM solution of NADH or NADPH has an apparent optical density of only 0.500.

A 0.1 mM solution of PEP has an absorbance of 0.167 at 240 nm. This wavelength is on the slope of the absorption curve, and erroneous results are obtained when too wide a slit is used. Using the Gilford Model 2400 spectrophotometer, an absorbance of 0.167 is obtained at slit widths of 0.8 mm (half-intensity bandwidth = 1.4 nm) or less. A slit width of 1.0 mm (half-intensity bandwidth = 1.8 nm) gives an extinction of 0.162, and a slit width of 1.5 mm (half-intensity bandwidth = 2.5 nm) an extinction of 0.144.

When absorbance measurements are made in the visible spectrum, it is much more convenient and less expensive to employ a colorimeter than to use a spectrophotometric colorimeter. We use a Coleman Jr. Model 1A instrument for measurements of hemoglobin concentration and glutathione. The largest cuvette for this instrument, having an internal diameter of 24 mm, has a critical volume of somewhat less than 10 ml and is used for the hemoglobin and glutathione determinations. For the reading of the Soret band of hemoglobin in estimating galactose-1-phosphate uridyl transferase (see Chapter 30), cuvettes with an outer diameter of 12 mm are used. Other relatively inexpensive colorimeters should be satisfactory as well. Naturally, any ultraviolet spectrophotometer can be used to make the same readings.

For the purposes outlined in this manual, a filter fluorometer is more satisfactory, and much less expensive, than a spectrofluorometer. We have found a Turner Model 110 or Model 111 to be very satisfactory. Inexpensive glass cuvettes may be used in all of the techniques described. The instrument should be fitted with a temperature control door, and the temperature of all reaction mixtures should be equilibrated at 37°, since the fluorescence of pyridine nucleotides is strongly temperature dependent.

The quantities of reagents to be diluted for the enzyme assays are generally quite small. For this reason, it is not desirable to use ordinary serologic pipets, even of the 0.1-ml variety, for making measurements. Glass micropipets, which are commercially available from several sources and which are highly accurate, are very satisfactory. However, such pipets are somewhat cumbersome to clean and are likely to produce considerable operator fatigue. When large numbers of enzyme assays are to be carried out, semiautomatic pipetting devices, such as that produced by Eppendorf, are extremely satisfactory. Although semiautomatic pipetting devices with volumes greater than 100 μl are also available, they seem to us to have no particular advantage over the use of ordinary straight serologic glass pipets.

The quantity of water used in most of the assay procedures has been calculated to give a final volume, for the complete system, of 1.000 ml. As a consequence, odd, and often difficult-to-measure quantities such as 605 or 715 μl are specified. The accuracy achieved with a straight 1-ml serologic pipet is sufficient for these measurements, however, since they affect only the final reaction volume, and a 10-μl error will only introduce a 1% error in the final concentration.

It is necessary to warm the contents of the cuvettes used for spectrophotometric assays prior to the initiation of the reaction. This can be accomplished merely by placing them in the thermostatically jacketed cuvette compartment of the spectrophotometer. However, this procedure makes this instrument unavailable for assays during the warmup incubation period. We have found the use of a temperature block, machined to fit the cuvette carrier

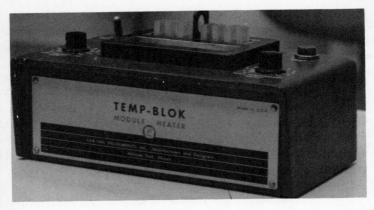

Fig. 4. A cuvette warming block used to prewarm cuvettes prior to assay. A block especially machined to hold the cuvette carrier is used. One cuvette carrier is in place; the second space is empty and could receive a second cuvette carrier.

(see Fig. 4), useful for the preincubation of spectrophotometric assay systems. Temperature block heaters are commercially available (e.g., Labline TempBlok), but it is necessary to custom-make the block insert for accommodating a cuvette carrier. Other equipment which is required for the satisfactory assay of red cell enzymes and metabolic intermediates includes a reliable pH meter, a high-quality analytical balance, and a refrigerated centrifuge.

6

calculation of results

A. METHODS OF EXPRESSING QUANTITY OF RED CELLS

The activity of red cell enzymes or levels of red cell intermediates may be expressed in three ways: in terms of the quantity (1) per cell, or, more conveniently, 10^{10} red cells, (2) per milliliter of red cells, or (3) per gram of hemoglobin. When the size or hemoglobin content of the red cells is abnormal, as is often the case in anemic subjects, interpretation of results may depend greatly on which means of expression is used. Thus, in a patient with the hypochromic microcytic anemia of iron deficiency, the activity of enzymes per gram of hemoglobin will usually be considerably increased but the activity per red cell may be diminished. The best means for the expression of red cell measurements has received detailed consideration [1].

In many ways, the most logical way of expressing the results of red cell measurements would be their concentration in the cell or in cell water. This would facilitate the use of results in kinetic analyses, and comparison with concentration in other tissues could be more easily carried out. But the difficulties in measuring the volume of red cells used in an estimation are greater than might appear at first sight. The washed packed red cells referred to in Chapter 3 are not, after all, packed red cells without the surrounding medium. They represent a suspension of cells in saline, and their actual volume could be approximated only by carrying out a hematocrit determination. The hematocrit of such packed red cells is generally found to be approximately 90%. Thus, the 1:20 hemolyzate referred to in Chapter 3 does not represent 50 μl red cells/ml hemolyzate. A second disadvantage of expressing results in terms of activity per milliliter of red cells is that errors in making dilution of hemolyzates contribute to the overall error inherent in the estimation. The pipetting of packed red cells, in particular, is fraught with error. This consideration is not a problem when activity is expressed in terms of grams of hemoglobin, since it is then possible to make the estimation of the quantity of red cells used on the final dilution of test material. The errors committed in all prior dilutions are therefore immaterial. The third important consideration is that the volume of red cells is altered by collection in ACD solution (they swell by approximately 20%) and may also change on storage in other preservative media. For these reasons, we have elected to rely primarily upon the concentrations of hemoglobin as a point of reference, although in a sense this is the least logical way of expressing the results of red cell enzyme assays and concentrations of red cell intermediates. Their association with the hemoglobin is, after all, a more or less accidental one. However, for the compelling technical reasons discussed above, we prefer to use this measurement.

The enumeration of red cells as a basis of measurement has the same technical advantages over the use of volume of red cells as does the quantitation of hemoglobin. However, the measurement of hemoglobin is carried out with much more dispatch in most laboratories where automatic counting equipment is unavailable than is the enumeration of red blood cells and the results have much higher accuracy.

When the red cells have a normal average size, hemoglobin content, and volume, all of these means of expressing concentration are essentially equiva-

Table V. Factors for Converting Enzyme Units Expressed in Terms of Different Red Cell Measurements

To convert	To	Multiply by
IU*/g Hb	IU/10^{10} RBC	0.01 × MCH ($\gamma\gamma$) (NV† = 0.29)
	IU/ml RBC	0.01 × MCHC (%) (NV = 0.34)
IU/ml RBC	IU/10^{10} RBC	0.01 × MCV (μ^3) (NV = 0.87)
	IU/g Hb	$\dfrac{100}{\text{MCHC (\%)}}$ (NV = 2.94)
IU/10^{10} RBC	IU/g Hb	$\dfrac{100}{\text{MCH } (\gamma\gamma)}$ (NV = 3.45)
	IU/ml RBC	$\dfrac{100}{\text{MCV } (\mu^3)}$ (NV = 1.15)

* International enzyme units.
† Normal value.

lent. Under these circumstances, simple conversion factors, given in Table V, may be employed.

In cases in which the red cells do not have a normal volume or hemoglobin content, conversion from quantity or activity per gram of hemoglobin to quantity or activity per milliliter of red cells or 10^{10} red cells can be accomplished with ease merely by performing a hematocrit, red blood cell count, and hemoglobin determination on the peripheral blood of the patient. The mean corpuscular hemoglobin concentration (MCHC), mean corpuscular volume (MCV), and mean corpuscular hemoglobin (MCH) are then calculated. The formula for converting enzyme units expressed in terms of one red cell measurement into another are given in Table V.

B. CALCULATION OF ENZYME ACTIVITY

In calculating the enzyme activity (E), in international units per gram of hemoglobin,

$$E = \frac{100A}{\text{Hb}} \tag{1}$$

where A is the number of enzyme units per milliliter and Hb is the concentration of hemoglobin in grams per 100 ml in the hemolyzate.

In the case of NADP- or NAD-linked enzyme assays in which 1 mole of coenzyme is reduced or oxidized for each mole of substrate consumed (e.g., triose phosphate isomerase, pyruvate kinase),

$$A = \frac{\Delta\text{OD}}{6.22} \times \frac{V_c}{V_H} \tag{2}$$

where V_c is the cuvette volume ($= 1.00$ in the systems described in Part III), V_H is the volume of hemolyzate in the reaction system, ΔOD is the change per minute in the optical density at 340 nm, and 6.22 is the optical density of a 1 mM solution of NAD(P)H.

In NADP- or NAD-linked enzyme assays in which 2 moles of coenzyme are reduced or oxidized for each mole of substrate consumed (e.g., phosphofructokinase or aldolase),

$$A = \frac{\Delta OD}{12.44} \times \frac{V_c}{V_H} \tag{3}$$

In calculating ΔOD from the progress curves, it is important to select the proper portion of the curve for measurement. In linked reactions the rate of change will often increase for several minutes and then become linear. It is this, most rapid, linear rate which is to be measured. On the other hand, as substrate is consumed, the rate of the enzymatic reaction will slow. Again, it is the most rapid portion of the progress curve which is to be measured. We have designed a very simple device for the measurement of the ΔOD. A piece of the recorder graph paper, corresponding in width to a 10-min reaction time and starting at the zero absorbance line is mounted between two pieces of clear plastic, of the type which is commonly employed for preserving identification cards and similar documents. Such plastic is readily available at stationery stores. One corner of the zero end of this measuring device is placed on a pencil line which has been fitted to the most rapid linear portion of the progress curve. The measuring device is aligned with the rectangular grid of the recorder paper, and the coordinate at which the line intersects the curve represents the change in optical density during a 10-min period (see Fig. 5). In calculating that ΔOD, it is important to remember not only to divide by the number of minutes over which the change of optical density has been measured (10 min if the measuring device referred to above is used) but also to take into account the full-scale setting of the recorder in optical density units (S). Any blank rate present in the absence of substrate is corrected for by reading the system against a no-substrate blank. Blank rates due to contaminating enzyme in the auxiliary enzyme preparations require separate correction by performing a blank assay in which β-mercaptoethanol-EDTA stabilizing solution is substituted for hemolyzate. This blank must be subtracted from the rate observed with hemolyzate. As indicated on page 11, 0.2 ml of 1:20 hemolyzate is added to 10 ml of ferricyanide-cyanide reagent. The optical density of this mixture is determined at 540 nm. The colorimeter should be periodically standardized with a commercially available cyanmethemoglobin standard solution (see page 11). The hemoglobin concentration of the 1:20 hemolyzate is generally approximately 1 g%. We generally do not make hemoglobin measurements on the 1:200 and 1:2000 dilutions, but

Fig. 5. A simple device for the measurement of slopes of assay curves. The device is prepared from a piece of recorder paper representing 10-min movement of the recorder chart paper, which has been placed between two clear plastic sheets of the type used to preserve identity cards and other documents. Such sheets are available at stationery shops. A line is fitted to the curve, and the lower left corner of the measuring device is placed on the line. The device is aligned with the grid of the corner paper, and the change in optical density for 10 min is read at the point where the line emerges from the measuring device.

merely calculate the hemoglobin concentration from the 1:20 dilution. Equations 1, 2, and 3 (pages 32 and 33), combined with the equation for Hb (page 12), may be generalized and the constant factors combined as follows to give the enzyme activity E in units per gram of Hb:

$$E = \frac{10(\Delta OD_R - \Delta OD_B) \times V_{Hb} \times S}{F_{HB} \times (V_{Fe} + V_{Hb}) \times OD_{540} \times \epsilon \times V_H}$$

where ΔOD_R is the apparent optical density change of the reaction system per 10 min; ΔOD_B is the apparent optical density change of the blank system for 10 min; V_{Hb} is the volume (in milliliters) of hemolyzate added to 10 ml of ferricyanide-cyanide reagent; F_{HB} is the factor required to convert to hemoglobin concentration in the cuvette to grams percent; V_{Fe} is the volume of ferricyanide-cyanide reagent; ϵ is the millimolar extinction coefficient (6.22 in reactions in which 1 mole of pyridine nucleotide is reduced or oxidized, 12.44 in reactions in which 2 moles of pyridine nucleotide are oxidized or reduced, and 13.6 when DTNB is reduced); V_H is the volume of hemolyzate added to a 1-ml mixture; S is the full-scale setting of the recorder; and OD_{540} is the optical density of the ferricyanide-cyanide reagent after the addition of hemolyzate.

C. CALCULATION OF CONCENTRATIONS OF INTERMEDIATES

Details of the calculation of concentrations of metabolic intermediates or coenzymes in red cells are presented with each individual method, since the techniques differ in so many respects. If the concentration is calculated from a known extinction coefficient of a light-absorbing reaction product, the millimolar concentration of the substance in the cuvette (C_s) is obtained from the equation

$$C_s = \frac{R_s}{\epsilon}$$

where ϵ is the optical density of a 1 mM solution of the reaction product and R_s is the reading obtained with the sample. To calculate the concentration of the substance in the original red cell sample, it is necessary to correct for the dilution occurring during the extraction process (see page 13) and for all subsequent dilutions. If the calculation is based on the reading obtained with a standard solution, the concentration of the substance in the sample (C_s) is obtained from the equation

$$C_s = \frac{R_s \times C_{st}}{R_{st}}$$

where R_{st} is the reading with a standard and C_{st} is the concentration of the standard. It is necessary to correct for those dilutions of the sample after the measurement of the concentration of red cells (by hemoglobin, hematocrit, or red count) but before the point in the procedure at which the sample and standard are treated in the same way.

In the case of fluorometric measurements of red cell intermediates the calculation of the concentration of the intermediates in the red cells is always based on the reading of a standard solution. Such calculations are intrinsic to fluorometric assays, since fluorescence is read in arbitrary units and depends not only on the intensity and spectral purity of the incident light but also on other factors such as temperature, quenching by unknown substances in the extract, and characteristics of the secondary filter. In the fluorometric assays, therefore, a standard solution, which has previously been assayed spectrophotometrically, is added to the assay system as the last step of the procedure. The change in fluorescence (F_{st}) produced by the standard represents the effect of oxidation or reduction of a stoichiometric amount of the pyridine nucleotide used in the assay and provides the basis of the calculation. In addition to the concentration (C_{st}) and volume (V_{st}) of the standard, the equation need only take into account the dilution of the red cell extract during preparation and in the assay system. The following equation may be used to calculate the concentration of intermediates by fluorometric pyridine nucleotide assays:

$$C_s = \frac{F_s \times C_{st} \times V_{st} \times V_{Neu}}{F_{st} \times V_p \times D \times V_s}$$

where V_{st} is the volume of the standard added to the assay system; V_p is the volume of PCA extract before neutralization; V_{Neu} is the volume of PCA extract after neutralization; V_s is the volume of the sample added to the assay system; and D is obtained from the equation on page 13.

REFERENCES

1. Hjelm, M.: The mode of expressing the content of intracellular components of human erythrocytes with special reference to adenine nucleotides. Scand. J. Haematol. 6:56–64, 1969.

part III

red cell glycolytic enzymes

7

hexokinase (Hx)

A. PRINCIPLE

Hexokinase catalyzes the reaction

$$\text{ATP} + \text{glucose} \xrightarrow[\text{Mg}^{2+}]{\text{Hx}} \text{Glucose-6-P} + \text{ADP}$$

In this assay the formation of glucose-6-P is measured by linking its further oxidation to 6-phosphogluconate to the reduction of NADP through the glucose-6-P dehydrogenase reaction:

$$\text{Glucose-6-P} + \text{NADP}^+ \xrightarrow{\text{G-6-PD}} \text{6-Phosphogluconate} + \text{NADPH} + \text{H}^+$$

Although 6-phosphogluconate dehydrogenase is present in the hemolyzate, we have found that the 6-phosphogluconate dehydrogenase formed is only partially converted to ribulose-5-phosphate in the 6-phosphogluconate dehydrogenase step.

B. PROCEDURE

The following reagents are added to cuvettes with a critical volume of less than 1 ml. If a large number of assays is to be carried out, multiples of the indicated volumes of tris buffer, glucose, ATP, NADP, MgCl₂, and water may be mixed and appropriate quantities of the mixture distributed into cuvettes.

	Blank (μl)	System (μl)	Low S system (μl)
Tris-HCl, 1 M, EDTA, 5 mM, pH 8.0	100	100	100
MgCl₂, 0.1 M	100	100	20
Glucose, 0.02 M	100	100	10
ATP (neut.), 0.02 M	—	500	50
NADP, 2 mM	100	100	100
1:20 hemolyzate	50	50	50
G-6-PD, 10 U/ml (diluted in β-mercaptoethanol-EDTA stabilizing solution)	10	10	10
H₂O	540	40	660

38

The increase of optical density of the system and low S system is measured against that of the blank at 340 nm at 37° C after preincubation for 15 min. A recorder expansion giving a full-scale reading of 0.2 OD unit is suitable for normal blood samples.

C. ADDITIONAL BLANKS REQUIRED

A blank assay should be carried out to be certain that the glucose-6-P dehydrogenase used is free of hexokinase activity. β-Mercaptoethanol-EDTA stabilizing solution is substituted for hemolyzate in both the blank and system mixtures, and the optical density of the hemolyzate-free system is measured against that of the hemolyzate-free blank for 10 min without preincubation. If no change in optical density is observed, indicating that the glucose-6-P dehydrogenase preparation is free of hexokinase, it is no longer necessary to carry out the blank determination each time. If some residual hexokinase activity is present in the glucose-6-P dehydrogenase, however, this blank should be determined each day. If it represents a rate of more than 0.0005 O.D. U/min, it is better to obtain a more purified preparation of glucose-6-P dehydrogenase.

If an appreciable blank rate is observed, a blank assay should also be carried out substituting β-mercaptoethanol-EDTA stabilizing solution for hemolyzate in the low S system. This blank should be used in computation of the activity of the low S system.

D. COMMENTS, CALCULATIONS, AND NORMAL VALUES

Hexokinase is the least active of the glycolytic enzymes. Good instrumentation is essential if meaningful results are to be obtained. The rate ordinarily does not become linear for some 15 to 20 min after the initiation of the reaction, and one should be certain that the maximum rate has been achieved before discontinuing the assay. The blank rate, if any (see Section C above) should be subtracted from the rate measured in the presence of hemolyzate. Exact computation of the rate of glucose phosphorylation is not possible because a small part of the 6-phosphogluconate formed in the glucose-6-phosphate dehydrogenase reaction is oxidized further to produce ribulose-5 phosphate and additional NADPH. On the other hand, some of the NADPH formed is consumed in reducing glucose to sorbitol in the L-hexonate dehydrogenase reaction, and possibly in other reactions as well. Measurement of the amount of ADP formed in the reaction suggests that the assumption that 1 mole of NADPH is formed for each mole of glucose phosphorylated is a relatively accurate one. For this reason, the equation on page 34 ($\epsilon = 6.22$) should be used to calculate the enzyme activity.

The red cells of normal adults contain 1.16 ± 0.17 IU of hexokinase/g Hb (mean $\pm$ standard deviation) at $37°$. The activity of the low S system is normally $50.5 \pm 5.38\%$ of that in the regular system at $37°$. Factors which permit conversion of results obtained at $25°$ and at $30°$ into values which would have been obtained with the same hemolyzate at $37°$ are presented in Appendix 2.

Young red cells have substantially more enzyme activity than do older cells: hexokinase is one of the enzymes in which the activity declines most rapidly during red cell aging. Hence, normal enzyme activity in the presence of reticulocytosis should be regarded with great suspicion. Several patients with hexokinase deficiency have been reported in the literature.

8

glucose phosphate isomerase (GPI or PHI)

A. PRINCIPLE

Glucose phosphate isomerase interconverts glucose-6-P and fructose-6-P:

$$\text{Glucose-6-P} \underset{\longleftarrow}{\overset{\text{GPI}}{\rightleftarrows}} \text{Fructose-6-P}$$

Fructose-6-P serves as a substrate in this assay, and the glucose-6-P formed is measured by linking it to the reduction of NADP through glucose-6-P dehydrogenase.

B. PROCEDURE

The following reagents are added to cuvettes with a critical volume of less than 1 ml. If a large number of assays is to be carried out, multiples of the indicated volumes of tris buffer, $MgCl_2$, NADP, fructose-6-P, and water may be mixed and appropriate quantities of the mixture distributed into cuvettes.

	Blank (μl)	System (μl)	Low S system (μl)
Tris-HCl, 1 M, EDTA, 5 mM, pH 8.0	100	100	100
MgCl$_2$, 0.1 M	100	100	100
NADP, 2 mM	100	100	100
Glucose-6-P dehydrogenase, 10 U/ml (diluted in β-mercaptoethanol-EDTA stabilizing solution)	10	10	10
Fructose-6-P, 0.02 M	—	100	5
H$_2$O	685	585	680

Incubate at 37° for 1 hour

| 1:20 hemolyzate | 5 | 5 | 5 |

The increase of optical density of the system and low S system is measured against that of the blank at 340 nm at 37° C for 10 to 20 min. A recorder expansion giving a full-scale reading of 1.0 OD unit is suitable for normal blood samples.

C. ADDITIONAL BLANKS REQUIRED

A blank assay should be carried out to be certain that the glucose-6-P dehydrogenase is free of GPI activity. This is done easily by measuring the optical density change of the reaction system as read against the blank system for 10 min prior to the initiation of the reaction by addition of hemolyzate. If no change in optical density is observed, indicating that the auxiliary enzyme preparation is free of GPI, it is not necessary to carry out the blank determination each time as long as the same auxiliary enzyme preparations are being used. If some residual GPI activity is present, however, this blank should be determined each day, and subtracted from the assay values.

D. COMMENTS, CALCULATIONS, AND NORMAL VALUES

A long period of preincubation without hemolyzate may be required because of a gradual increase in optical density which sometimes initially occurs in this system, apparently independently of any GPI activity. This is probably due to the gradual oxidation of glucose-6-P contaminating fructose-6-P. Although with some preparations a full hour's incubation is necessary to reduce this nonspecific rate to negligible levels, this may not always be the case. In any event, any optical density change of the system without hemolyzate should be minimal at the time that the reaction is started by the addition of hemolyzate.

The blank rate, if any (see Section C above), should be subtracted from the rate measured in the presence of hemolyzate. For reasons that are not entirely clear, only 1 mole of NADP is reduced for each mole of fructose-6-

phosphate converted to glucose-6-phosphate. 6-PGA accumulates, and the rate of the 6-phosphogluconic dehydrogenase reaction appears to be negligible in this assay. The equation on page 34 ($\epsilon = 6.22$) is used to calculate the enzyme activity.

The red cells of normal adults contain 60.8 ± 11.0 EU of GPI/g Hb (mean $\pm$ standard deviation) at $37°$. The activity of the low S system is normally $46.2 \pm 2.41\%$ of that in the regular system at $37°$. Factors which permit conversion of results obtained at $25°$ and at $30°$ into values which would have been obtained with the same hemolyzate at $37°$ are presented in Appendix 2. Several cases of GPI deficiency resulting in nonspherocytic congenital hemolytic anemia have been reported to occur.

9

phosphofructokinase (PFK)

A. PRINCIPLE

Phosphofructokinase catalyzes the phosphorylation of fructose-6-P by ATP to fructose-1,6-diP:

$$\text{Fructose-6-P} + \text{ATP} \xrightarrow[\text{Mg}^{2+}]{\text{PFK}} \text{Fructose-1,6-diP} + \text{ADP}$$

In this assay the fructose-1,6-diP formed is measured by conversion to dihydroxyacetone phosphate (DHAP) through the aldolase and triose phosphate isomerase (TPI) reactions. The dihydroxyacetone phosphate is then reduced by the action of α-glycerophosphate dehydrogenase, oxidizing NADH to NAD. The sequence of reactions involved is as follows:

$$\text{Fructose-1,6-diP} \xrightleftharpoons{\text{aldolase}} \text{Glyceraldehyde-3-P} + \text{DHAP}$$

$$\text{Glyceraldehyde phosphate} \xrightleftharpoons{\text{TPI}} \text{DHAP}$$

$$\text{DHAP} + \text{NADH} + \text{H}^+ \xrightarrow{\substack{\alpha\text{-glycerophosphate} \\ \text{dehydrogenase}}} \alpha\text{-Glycerophosphate} + \text{NAD}^+$$

The oxidation of NADH is measured at 340 nm.

Phosphofructokinase is an allosteric enzyme which is strongly stimulated by ADP when the enzyme is not saturated with fructose-6-phosphate. The capacity of the enzyme to respond to ADP is tested in the "low S + ADP" system.

B. PROCEDURE

The following reagents are added to cuvettes with a critical volume of less than 1 ml. If a large number of assays is to be carried out, multiples of the indicated volumes of tris buffer, ATP, $MgCl_2$, fructose-6-P, NADH, and water may be mixed and appropriate quantities of the mixture distributed into cuvettes.

	Blank (μl)	System (μl)	Low S system (μl)	Low S + ADP (μl)
Tris-HCl, 1 M, EDTA, 5 mM, pH 8.0	100	100	100	100
ADP (neut.), 0.03 M	—	—	—	10
$MgCl_2$, 0.1 M	200	200	200	200
Fructose-6-P, 0.02 M	—	100	5	5
NADH, 2 mM	100	100	100	100
Auxiliary enzyme solution*	100	100	100	100
1:20 hemolyzate	10	10	10	10
H_2O	390	290	385	375
Incubate at 37° for 10 min				
ATP (neut.), 0.02 M	100	100	100	100

* Auxiliary enzyme solution: Make $(NH_4)_2SO_4$ suspension of 50 units of each enzyme per milliliter by adding 100 units of aldolase, 100 units of triose phosphate isomerase, and 100 units of α-glycerophosphate dehydrogenase to a tube and bringing to a total volume of 2 ml with saturated $(NH_4)_2SO_4$ solution. The suspension is stable at 4° C. Dilute 1:15 in β-mercaptoethanol-EDTA stabilizing solution.

A recorder expansion giving a full-scale reading of 0.4 OD units is suitable for normal blood samples. When using a recording spectrophotometer which does not have a provision for arbitrarily adjusting the position of the recorder pen, or when using a nonrecording spectrophotometer, the instrument should be blanked at an optical density value of 0.3, since the optical density of the system will fall.

C. ADDITIONAL BLANKS REQUIRED

A blank assay should be carried out to be certain that the aldolase, TPI, and α-glycerophosphate dehydrogenase used are free of PFK activity. β-Mercaptoethanol-EDTA stabilizing solution is substituted for hemolyzate in both the blank and system mixtures, and the change in optical density of the hemo-lyzate-free system is measured against that of the hemolyzate-free blank. If no change in optical density is observed, indicating that the auxiliary enzyme preparation is free of PFK, it is only necessary to carry out the blank determi-nation each time a new auxiliary enzyme preparation is made. If some residual PFK activity is present in auxiliary enzymes, however, this blank rate should be determined each day.

The source of contamination is sometimes difficult to identify if no uncon-taminated enzyme preparations are available. The best way to detect the source of contaminating PFK activity is to prepare three new enzyme suspen-sions: one containing twice the amount of TPI, one containing twice the amount of α-glycerophosphate dehydrogenase, and one containing twice the amount of aldolase specified. Increasing the amount of each of these auxiliary enzymes should not appreciably affect the rate of the blank reaction, unless the enzyme is contaminated with PFK.

If an appreciable blank rate is observed, a blank assay should also be car-ried out substituting β-mercaptoethanol-EDTA stabilizing solution for hemo-lyzate in the low S system. This blank should be used in computation of the activity of the low S system.

D. COMMENTS, CALCULATIONS, AND NORMAL VALUES

PFK is less stable in hemolyzates than are the other glycolytic enzymes. About 17% of activity is lost when the hemolyzate is stored at 0° for 5 hours. As activity of the enzyme is lost, the activity in the low S system decreases to a lesser extent than does the activity in the system with high substrate con-centrations. This is probably the case because red cells contain two phospho-fructokinase isozymes, and the K_m for fructose-6-P of the unstable component is apparently higher than that of the stable component.

In the low S system, a rapid initial rate is observed if the sample is not preincubated for 10 min before the reaction is initiated with ATP. This rate is presumably due to the fact that the fructose-6-P concentration falls rapidly as fructose-6-P is converted to an equilibrium mixture of glucose-6-P and fructose-6-P in the presence of glucose phosphate isomerase. Since the con-centration of fructose-6-P limits the rate of the reaction in the low S system, the reaction rate will fall until the equilibrium concentration of fructose-6-P has been reached. The blank rate, if any (see Section C above), should be subtracted from the rate measured in the presence of hemolyzate. Two moles of NADH are oxidized for each mole of fructose-6-phosphate phosphorylated. The equation on page 34 ($\epsilon = 12.44$) is used to calculate the enzyme activity.

The red cells of normal adults contain 11.01 ± 2.33 IU of PFK/g Hb (mean ± standard deviation) at 37°. The activity of the low S system is normally 16.5 ± 2.01% and that of the low S + ADP system 29.9 ± 4.15% of that in the regular system at 37°. Factors which permit conversion of results obtained at 25° and at 30° into values which would have been obtained with the same hemolyzate at 37° are presented in Appendix 2.

10

aldolase

A. PRINCIPLE

Aldolase catalyzes the cleavage of fructose-1,6-diP into two molecules of triose phosphate, glyceraldehyde-3-phosphate (GAP), and dihydroxyacetone phosphate (DHAP):

$$\text{Fructose-1,6-diP} \underset{\text{}}{\overset{\text{aldolase}}{\rightleftharpoons}} \text{GAP} + \text{DHAP}$$

The amount of GAP and DHAP formed is measured by converting GAP to DHAP through the TPI reaction and then reducing the DHAP formed through the α-glycerophosphate dehydrogenase reaction:

$$\text{DHAP} + \text{NADH} + \text{H}^+ \rightleftharpoons \alpha\text{-Glycerophosphate} + \text{NAD}^+$$

The oxidation of NADH to NAD in this process is measured at 340 nm.

B. PROCEDURE

The following reagents are added to cuvettes with a critical volume of less than 1 ml. If a large number of assays is to be carried out, multiples of the indicated volumes of tris buffer, NADH, and water may be mixed and appropriate quantities of the mixture distributed into cuvettes.

	Blank (μl)	System (μl)	Low S system (μl)
Tris-HCl, 1 M, EDTA, 5 mM, pH 8.0	100	100	100
NADH, 2 mM	100	100	100
Auxiliary enzyme solution*	100	100	100
1:20 hemolyzate†	10	10	10
H$_2$O	690	590	680

Incubate at 37° C for 10 min

Fructose-1-6-diP, 0.01 M	—	100	10

* Auxiliary enzyme solution: Make an $(NH_4)_2SO_4$ suspension of 50 U/ml of each enzyme by adding 100 units of TPI and 100 units of α-glycerophosphate dehydrogenase to a tube and bringing to a volume of 2 ml with saturated $(NH_4)_2SO_4$ solution. The suspension is stable at 4° C. For assay dilute 1:15 in β-mercaptoethanol-EDTA stabilizing solution.

† Stored in ice for 2 to 6 hours prior to assay.

The decrease of optical density of the system of low S system is measured against that of the blank at 340 nm at 37° C for 10 to 20 min.

A recorder expansion giving a full-scale reading of 0.2 OD unit is suitable for normal blood samples. When using a recording spectrophotometer which does not have a provision for arbitrarily adjusting the position of the recorder pen, or when using a nonrecording spectrophotometer, the instrument should be blanked at an optical density value of 0.1, since the optical density of the system will fall.

C. ADDITIONAL BLANKS REQUIRED

A blank assay should be carried out to be certain that the TPI and α-glycerophosphate dehydrogenase used are free of aldolase activity. β-Mercaptoethanol-EDTA stabilizing solution is substituted for hemolyzate in both the blank and system mixtures, and the optical density of the hemolyzate-free system is measured against that of the hemolyzate-free blank. If no change in optical density is observed, indicating that the auxiliary enzyme preparation is free of aldolase activity, it is no longer necessary to carry out the blank determination each time. If some residual aldolase activity is present in the auxiliary enzymes, however, this blank should be determined each day.

If an appreciable blank rate is observed, a blank assay should also be carried out substituting β-mercaptoethanol-EDTA stabilizing solution for hemolyzate in the low S system. This blank should be used in computation of the activity of the low S system.

D. COMMENTS, CALCULATIONS, AND NORMAL VALUES

Because of the low activity of this enzyme, instrumentation of high sensitivity is required for satisfactory results. There is considerable binding of aldolase by red cell stroma, and it is imperative, for this reason, that uncentrifuged hemolyzate be assayed.

Freezing and thawing of the hemolyzate increases its enzyme activity. Repeated freezing and thawing does not result in further increase in activity, and detergents such as Triton X-100 and digitonin were not found to activate the enzyme. There seems to be a slight increase in the activity of the enzyme for about 2 hours after the preparation of the freeze-thaw hemolyzate. After this, the enzyme activity appears to be relatively stable.

The blank rate, if any (see Section C above), should be subtracted from the rate measured in the presence of hemolyzate. Two moles of NADH are oxidized for each mole of fructose-1,6-diphosphate converted to triose phosphate. The equation on page 34 ($\epsilon = 12.44$) is used to calculate the enzyme activity. The red cells of normal adults contain 3.19 ± 0.86 IU of aldolase/g Hb (mean $\pm$ standard deviation) at 37°. The activity of the low S system is normally $64.1 \pm 7.40\%$ of that in the regular system at 37°. Factors which permit conversion of results obtained at 25° and at 30° into values which would have been obtained with the same hemolyzate at 37° are presented in Appendix 2. Aldolase activity is markedly influenced by red cell age, and the activity of this enzyme is increased in patients with hemolytic disease. A single case of aldolase deficiency resulting in nonspherocytic congenital hemolytic anemia and mental retardation has been detected.

11

triose phosphate isomerase (TPI)

A. PRINCIPLE

Triose phosphate isomerase interconverts glyceraldehyde-3-P (GAP) and dihydroxyacetone phosphate (DHAP):

$$GAP \overset{TPI}{\rightleftharpoons} DHAP$$

The rate of DHAP formation from GAP is measured by linking it to the oxidation of NADH to NAD through the α-glycerophosphate dehydrogenase reaction:

$$\text{NADH} + \text{H}^+ + \text{DHAP} \xrightleftharpoons[\text{dehydrogenase}]{\alpha\text{-glycerophosphate}} \text{NAD}^+ + \alpha\text{-glycerophosphate}$$

The oxidation of NADH is measured at 340 nm.

B. PROCEDURE

The following reagents are added to cuvettes with a critical volume of less than 1 ml. If a large number of assays is to be carried out, multiples of the indicated volumes of tris buffer, NADH, and water may be mixed and appropriate quantities of the mixture distributed into cuvettes.

	Blank (μl)	System (μl)
Tris-HCl, 1 M, EDTA, 5 mM, pH 8.0	100	100
NADH, 2 mM	100	100
1:2000 hemolyzate	10	10
α-Glycerophosphate dehydrogenase, $\simeq$2 U/ml	50	50
H$_2$O	740	640

Incubate at 37° for 10 min

D-GAP,* 30 mM	—	100

* To adjust the concentration of DL-GAP to 30 mM D-GAP, dilute commercial DL-GAP (50 mg/ml, Sigma) 1:10 and assay in the following system:

	Blank (μl)	System (μl)
Tris-HCl, 1 M, EDTA, 5 mM, pH 8.0	100	100
NADH, 2 mM	100	100
TPI (undiluted suspension)	5	5
α-Glycerophosphate dehydrogenase, $\simeq$2 U/ml	50	50
H$_2$O	740	740

Read baseline OD

DL-GAP (1:10)	—	5
H$_2$O	5	—

Read at 340 nm until no further change in OD occurs. The concentration of D-GAP in the stock solution (mM) is

$$C = \frac{\Delta\text{OD} \times 2000}{6.22}$$

Take 0.30 ml undiluted DL-GAP, and bring volume to 0.01C ml with H$_2$O to obtain a solution containing 30 mM D-GAP.

The decrease of the optical density of the system is measured against that of the blank at 340 mm at 37° C for approximately 10 min.

A recorder expansion giving a full-scale reading of 0.4 OD unit is suitable for normal blood samples. When using a recording spectrophotometer which does not have a provision for arbitrarily adjusting the position of the recorder pen, or when using a nonrecording spectrophotometer, the instrument should be blanked at an optical density value of 0.3, since the optical density of the system will fall.

C. ADDITIONAL BLANKS REQUIRED

A blank assay should be carried out to be certain that the α-glycerophosphate dehydrogenase used is free of TPI activity. β-Mercaptoethanol-EDTA stabilizing solution is substituted for hemolyzate in both the blank and system mixtures, and the optical density of the hemolyzate-free system is measured against that of the hemolyzate-free blank. If no change in optical density is observed, indicating that the α-glycerophosphate dehydrogenase preparation is free of TPI, it is not necessary to carry out the blank determination each time as long as the same α-glycerophosphate dehydrogenase preparation is used. If some residual TPI activity is present in the glycerophosphate dehydrogenase, however, this blank should be determined each day.

D. COMMENTS, CALCULATIONS, AND NORMAL VALUES

The reaction rate is very strongly dependent upon the concentration of D-GAP. Thus, the concentration of substrate is quite critical. It is for this reason that it is useful to assay all new lots of DL-GAP to determine their true D-GAP content, and to dilute them to give the 30 mM concentration required. It is also noteworthy that D-GAP exists in two configurational forms, one of which is rapidly metabolized and one of which is slowly metabolized. Therefore, the reaction will have an initial rapid phase, which should be measured in calculating the results, and a slower phase, which may be ignored.

Another factor which results in variability in the results of this assay is the fact that triose phosphate isomerase is markedly inhibited by ammonium sulfate. Since α-glycerophosphate dehydrogenase is sold as an ammonium sulfate suspension, appreciable amounts of ammonium sulfate solution are introduced into the assay mixture with the addition of this auxiliary enzyme. The amount of ammonium sulfate will depend on the specific activity of the particular lot of enzyme purchased. Relatively little inhibition will be observed when the α-glycerophosphate dehydrogenase is of high activity (600 U/ml or higher). As little as 0.5 mM ammonium sulfate in the final suspension will produce approximately 10% inhibition of TPI activity.

The blank rate, if any (see Section C above), should be subtracted from the rate measured in the presence of hemolyzate. One mole of NADH is oxidized for each mole of D-GAP converted to DHAP. The equation on page 34 ($\epsilon = 6.22$) is used to calculate the enzyme activity. The red cells of normal adults contain 2111 ± 397 IU of TPI/g Hb (mean $\pm$ standard deviation) at $37°$. Factors which permit conversion of result obtained at $25°$ and at $30°$ into values which would have been obtained with the same hemolyzate at $37°$ are presented in Appendix 2. Several cases of TPI deficiency resulting in nonspherocytic congenital hemolytic anemia have been reported in the literature.

12

glyceraldehyde phosphate dehydrogenase (GAPD)

A. PRINCIPLE

Glyceraldehyde phosphate dehydrogenase catalyzes the phosphorylation and oxidation of glyceraldehyde-3-phosphate (GAP) to 1,3-diphosphoglycerate (1,3-DPG):

$$GAP + P_i + NAD^+ \overset{GAPD}{\rightleftharpoons} 1,3\text{-}DPG + NADH + H^+$$

The reaction is run from the right to the left. Since 1,3-DPG is very unstable, and therefore not commercially available, it is generated from 3-PGA in the phosphoglycerate kinase (PGK) reaction:

$$3\text{-}PGA + ATP \overset{PGK}{\rightleftharpoons} 1,3\text{-}DPG + ADP$$

The oxidation of NADH is measured at 340 nm.

B. PROCEDURE

The following reagents are added to cuvettes with a critical volume of less than 1 ml. If a large number of assays is to be carried out, multiples of the indicated volumes of tris buffer, NADH, MgCl$_2$, ATP, PGK, and water may be mixed and appropriate quantities of the mixture distributed into cuvettes.

	Blank (μl)	System (μl)
Tris-HCl, 1 M, EDTA, 5 mM, pH 8.0	100	100
MgCl$_2$, 0.1 M	100	100
NADH, 2 mM	100	100
ATP (neut.) 20 mM	400	400
PGK, 50 U/ml	100	100
1:200 hemolyzate	10	10
H$_2$O	190	90

Incubate at 37° C for 10 min

| 3-PGA, 100 mM | — | 100 |

The decrease of optical density of the system is measured against that of the blank at 340 nm at 37° C for 10 to 20 min. A recorder expansion giving a full-scale reading of 1.0 OD unit is suitable for normal blood samples. When using a recording spectrophotometer which does not have a provision for arbitrarily adjusting the position of the recorder pen, or when using a nonrecording spectrophotometer, the instrument should be blanked at an optical density value of 0.6, since the optical density of the system will fall.

C. ADDITIONAL BLANKS REQUIRED

A blank assay should be carried out to be certain that the PGK used is free of GAPD activity. β-Mercaptoethanol-EDTA stabilizing solution is substituted for hemolyzate in both the blank and system mixtures, and the optical density of the hemolyzate-free system is measured against that of the hemolyzate-free blank. If no change in optical density is observed, indicating that the PGK is free of GAPD activity, it is no longer necessary to carry out the blank determination each time. If, however, some residual GAPD activity is present in the auxiliary enzyme, this blank should be determined each day.

D. COMMENTS, CALCULATIONS, AND NORMAL VALUES

This enzyme is strongly bound to stroma, the degree of binding apparently being much greater at a slightly alkaline than at an acid pH. It is essential, therefore, to carry out the assay on an uncentrifuged hemolyzate. The blank rate, if any (see Section C above), should be subtracted from the rate measured in the presence of hemolyzate. One mole of NADH is oxidized for each mole of 1,3-DPG converted to GAP. The equation on page 34 ($\epsilon =$ 6.22) is used to calculate the enzyme activity.

The red cells of normal adults contain 226 ± 41.9 IU of GAPD/g Hb (mean $\pm$ standard deviation) at 37°. Factors which permit conversion of results obtained at 25° and at 30° into values which would have been obtained with the same hemolyzate at 37° are presented in Appendix 2.

13

phosphoglycerate kinase (PGK)

A. PRINCIPLE

Phosphoglycerate kinase catalyzes the phosphorylation of ADP to ATP by 1,3-diphosphoglycerate (1,3-DPG):

$$\text{1,3-DPG} + \text{ADP} \overset{\text{PGK}}{\rightleftharpoons} \text{3PGA} + \text{ATP}$$

In the assay procedure, the reaction is measured in the reverse direction (from right to left). The formation of 1,3-DPG is then measured through the glyceraldehyde phosphate dehydrogenase (GAPD) reaction:

$$\text{GAP} + \text{NAD}^+ + \text{P}_i \overset{\text{GAPD}}{\rightleftharpoons} \text{NADH} + \text{H}^+ + \text{1,3-DPG}$$

in the reverse (right to left) direction. The oxidation of NADH is measured at 340 nm.

B. PROCEDURE

The following reagents are added to cuvettes with a critical volume of less than 1 ml. If a large number of assays is to be carried out, multiples of the indicated volumes of tris buffer, $MgCl_2$, ATP, NADH, and water may be mixed and appropriate quantities of the mixture distributed into cuvettes.

	Blank (μl)	System (μl)	Low S system (μl)
Tris-HCl, 1 M, EDTA, 5 mM, pH 8.0	100	100	100
MgCl$_2$, 0.1 M	100	100	100
ATP (neut.), 0.02 M	400	400	400
GAPD, 40 U/ml (diluted in β-mercaptoethanol-EDTA stabilizing solution)	100	100	100
NADH, 2 mM	100	100	100
1:200 hemolyzate)	10	10	10
H$_2$O	190	90	185
Incubate at 37° C for 10 min			
3-PGA, 100 mM	—	100	5

The decrease of optical density of the system is measured against that of the blank at 340 nm at 37° C for 10 to 20 min.

A recorder expansion giving a full-scale reading of 1.0 OD unit is suitable for normal blood samples. When using a recording spectrophotometer which does not have a provision for arbitrarily adjusting the position of the recorder pen, or when using a nonrecording spectrophotometer, the instrument should be blanked at an optical density value of 0.6, since the optical density of the system will fall.

C. ADDITIONAL BLANKS REQUIRED

A blank assay should be carried out to be certain that the GAPD used is free of PGK activity. β-Mercaptoethanol-EDTA stabilizing solution is substituted for hemolyzate in both the blank and system mixtures, and the optical density of the hemolyzate-free system is measured against that of the hemolyzate-free blank for 45 min without preincubation. If no change in optical density is observed, indicating that the GAPD preparation is free of PGK, it is no longer necessary to carry out the blank determination each time. If some residual PGK activity is present in the GAPD, however, this blank should be determined each day.

D. COMMENTS, CALCULATIONS, AND NORMAL VALUES

Some batches of GAPD are quite unstable and should be assayed periodically diluted in β-mercaptoethanol-EDTA stabilizing solution in the system described in Chapter 12 to be certain that adequate potency is present. The blank rate, if any (see Section C above), should be subtracted from the rate measured in the presence of hemolyzate. One mole of NADH is oxidized for each mole of 3-PGA converted to 1,3-DPG. The equation on page 34 ($\epsilon = 6.22$) is used to calculate the enzyme activity.

The red cells of normal adults contain 320 ± 36.1 IU of PGK/g Hb (mean $\pm$ standard deviation) $37°$. The activity of the low S system is normally $56.2 \pm 5.33\%$ of that in the regular system at $37°$. Factors which permit conversion of results obtained at $25°$ and at $30°$ into values which would have been obtained with the same hemolyzate at $37°$ are presented in Appendix 2.

14

diphosphoglycerate mutase (DPGM)

A. PRINCIPLE

Diphosphoglycerate mutase catalyzes the conversion of 1,3-DPG to 2,3-DPG:

$$1,3\text{-DPG} \xrightarrow[\text{3-PGA}]{\text{DPGM}} 2,3\text{-DPG}$$

In this assay, 1,3-DPG is formed by incubating F-1,6-diP with aldolase, TPI, and GAPD through a series of reactions:

$$\text{F-1,6-diP} \xrightarrow{\text{aldolase}} \text{GAP} + \text{DHAP}$$
$$\text{GAP} \xrightleftharpoons{\text{TPI}} \text{DHAP}$$
$$\text{P}_i + \text{NAD} + \text{GAP} \xrightarrow{\text{GAPD}} 1,3\text{-DPG} + \text{NADH}$$

An equilibrium mixture of F-1,6-diP, GAP, DHAP, 1,3-DPG, NAD, and NADH is formed. Hemolyzate is then added. Since very little ADP is present, PGK cannot function; therefore the removal of 1,3-DPG is almost entirely a function of DPGM activity. As the GAP/1,3 DPG equilibrium is displaced, NAD is reduced to NADH. The reduction of NAD is measured at 340 nm.

B. PROCEDURE (based on Schröter and Kalinowsky [1])

The following reagents are added to cuvettes with a critical volume of less than 1 ml. If a large number of assays is to be carried out, multiples of the indicated volumes of tris buffer, NAD, F-1,6-diP, 3-PGA, KH_2PO_4, and water may be mixed and appropriate quantities of the mixture distributed into cuvettes.

	Blank (μl)	System (μl)
Tris-HCl, 1 M, EDTA, 5 mM, pH 8.0	100	100
NAD, 10 mM	100	100
F-1, 6-diP, 100 mM	—	50
3-PGA, 10 mM	100	100
KH_2PO_4, 100 mM	20	20
GAPD, 50 U/ml (diluted in β-mercaptoethanol-EDTA stabilizing solution)	20	20
Aldolase, 5 U/ml	20	20
TPI, 60 U/ml	20	20
H_2O	600	550

Incubate at 37° C for 10 min

1:20 hemolyzate	20	20

After addition of the hemolyzate the increase in the optical density of the system is measured against that of the blank for 10 to 15 min.

A recorder expansion giving a full-scale reading of 0.4 OD unit is suitable for normal blood samples.

C. ADDITIONAL BLANKS REQUIRED

A blank assay should be carried out to be certain that the aldolase, TPI, and GAPD used are free of DPGM activity. β-Mercaptoethanol-EDTA stabilizing solution is substituted for hemolyzate in both the blank and system mixtures, and the optical density of the hemolyzate-free system is measured against that of the hemolyzate-free blank. If no change in optical density is observed, indicating that the auxiliary enzymes are free of DPGM activity, it is no longer necessary to carry out a blank determination each time. If some residual DPGM activity is present in the auxiliary enzymes, this blank should be determined each day.

D. COMMENTS, CALCULATIONS, AND NORMAL VALUES

The blank rate, if any (see Section C above), should be subtracted from the rate measured in the presence of hemolyzate. One mole of NAD is reduced for each mole of 1,3-DPG converted 2,3-DPG. The equation on page 34 ($\epsilon = 6.22$) is used to calculate the enzyme activity.

The red cells of normal adults contain 4.78 ± 0.65 IU of DPGM/g Hb (mean $\pm$ standard deviation) at 37°. Factors which permit conversion of results obtained at 25° and at 30° into values which would have been obtained with the same hemolyzate at 37° are presented in Appendix 2.

REFERENCE

1. Schröter, W., and Kalinowsky, W.: Erythrocyte 2,3-diphosphoglycerate mutase: an optical test in hemolysates. Clin. Chim. Acta 26:283–285, 1969.

15

monophosphogylceromutase (MPGM)

A. PRINCIPLE

Monophosphoglyceromutase catalyzes the equilibrium between 3-phosphoglyceric acid (3-PGA) and 2-phosphoglyceric acid (2-PGA):

$$3\text{-PGA} \underset{\text{2,3-DPG}}{\overset{\text{MPGM}}{\rightleftharpoons}} 2\text{-PGA}$$

The assay is carried out using 3-PGA as substrate. 2,3-DPG is an essential cofactor. The 2-PGA formed is converted to phosphoenolpyruvate (PEP) through the enolase reaction:

$$2\text{-PGA} \overset{\text{enolase}}{\rightleftharpoons} \text{PEP}$$

The formation of PEP is measured by converting it to pyruvate in the pyruvate kinase (PK) reaction:

$$\text{PEP} + \text{ADP} \overset{\text{PK}}{\rightarrow} \text{Pyruvate} + \text{ATP}$$

The pyruvate oxidizes NADH in the lactate dehydrogenase (LDH) reaction:

$$\text{Pyruvate} + \text{NADH} + \text{H}^+ \overset{\text{LDH}}{\rightleftharpoons} \text{Lactate} + \text{NAD}^+$$

The oxidation of NADH to NAD is followed at 340 nm.

B. PROCEDURE

The following reagents are added to cuvettes with a critical volume of less than 1 ml. If a large number of assays is to be carried out, multiples of the indicated volumes of tris buffer, 2,3-DPG, $MgCl_2$, KCl, NADH, ADP, LDH, PK, enolase, and water may be mixed and appropriate quantities of the mixture distributed into cuvettes.

	Blank (μl)	System (μl)	Low S system (μl)
Tris-HCl, 1 M, EDTA, 5 mM, pH 8.0	100	100	100
$MgCl_2$, 0.1 M	100	100	100
KCl, 1.0 M	100	100	100
NADH, 2 mM	100	100	100
ADP (neut.), 30 mM	50	50	50
2,3-DPG, 1 mM	10	10	10
LDH, 60 U/ml	10	10	10
PK, 50 U/ml	10	10	10
Enolase, 10 U/ml	10	10	10
1:20 hemolyzate	20	20	20
H_2O	490	290	470
Incubate at 37° C for 10 min			
3-PGA, 10 mM	—	200	20

The decrease of optical density of the system and low S system is measured against that of the blank at 340 nm at 37° C for 10 to 20 min. A recorder expansion giving a full-scale reading of 0.4 OD unit is suitable for normal blood samples. When using a spectrophotometer which does not have the provision for arbitrarily adjusting the position of the recorder pen, or when using a nonrecording spectrophotometer, the instrument should be blanked at an optical density value of 0.6, since the optical density of the system will fall.

C. ADDITIONAL BLANKS REQUIRED

A blank assay should be carried out to be certain that the enolase, PK, and LDH are free of MPGM activity. β-Mercaptoethanol-EDTA stabilizing solution is substituted for the hemolyzate in both the blank and system mixtures, and the optical density of the hemolyzate-free system is measured against that of the hemolyzate-free blank. If no change in optical density is observed, indicating that the enolase, PK, and LDH are free of MPGM activity, it is not necessary to carry out the blank determination each time. If some residual MPGM activity is present, however, this blank should be determined each day.

If an appreciable blank rate is observed, a blank assay should also be carried out substituting β-mercaptoethanol-EDTA stabilizing solution for hemolyzate in the low S system. This blank should be used in computation of the activity of the low S system.

D. COMMENTS, CALCULATIONS, AND NORMAL VALUES

One mole of PEP is formed for each mole of 2-PGA utilized. This results, ultimately, in the oxidation of 1 mole of NADH. The equation on page 34 ($\epsilon = 6.22$) is used to calculate the enzyme activity.

The red cells of normal adults contain 19.3 ± 3.84 IU of MPGM/g Hb (mean $\pm$ standard deviation) at $37°$. The activity of the low S system is normally $49.8 \pm 5.48\%$ of that in the regular system at $37°$. Factors which permit conversion of results obtained at $25°$ and at $30°$ into values which would have been obtained with the same hemolyzate at $37°$ are presented in Appendix 2.

16

enolase

A. PRINCIPLE

Enolase catalyzes equilibrium between 2-phosphoglyceric acid (2-PGA) and phosphoenolpyruvic acid (PEP):

$$2\text{-PGA} \xrightleftharpoons{\text{enolase}} \text{PEP}$$

The formation of PEP is measured by converting it to pyruvate in the pyruvate kinase reaction:

$$\text{PEP} + \text{ADP} \xrightarrow{\text{PK}} \text{Pyruvate} + \text{ATP}$$

The pyruvate formed oxidizes NADH in the lactic dehydrogenase (LDH) reaction:

$$\text{Pyruvate} + \text{NADH} + \text{H}^+ \overset{\text{LDH}}{\rightleftharpoons} \text{Lactate} + \text{NAD}^+$$

The oxidation of NADH to NAD is measured at 340 nm.

B. PROCEDURE

The following reagents are added to cuvettes with a critical volume of less than 1 ml. If a large number of assays is to be carried out, multiples of the indicated volumes of tris buffer, $MgCl_2$, KCl, NADH, ADP, LDH, PK, and water may be mixed and appropriate quantities of the mixture distributed into cuvettes.

	Blank (μl)	System (μl)	Low S system (μl)
Tris-HCl, 1 M, EDTA, 5 mM, pH 8.0	100	100	100
$MgCl_2$, 0.1 M	100	100	100
KCl, 1 M	100	100	100
NADH, 2 mM	100	100	100
ADP, 30 mM	50	50	50
LDH, 60 U/ml	10	10	10
PK, 50 U/ml	10	10	10
1:20 hemolyzate	20	20	20
H_2O	510	410	505

Incubate at 37° C for 10 min

2-PGA, 10 mM	—	100	5

The increase of optical density of the system and low S system is measured against that of the blank at 340 nm at 37° C. A recorder expansion giving a full-scale reading of 0.1 OD unit is suitable for normal blood samples. When using a spectrophotometer which does not have the provision for arbitrarily adjusting the position of the recorder pen, or when using a nonrecording spectrophotometer, the instrument should be blanked at an optical density value of 0.06, since the optical density of the system will fall.

C. ADDITIONAL BLANKS REQUIRED

A blank assay should be carried out to make certain that the PK and LDH used are free of enolase activity. β-Mercaptoethanol-EDTA stabilizing solution is substituted for the hemolyzate in both the blank and system mixtures, and the optical density of the hemolyzate-free system is measured against that of the hemolyzate-free blank. If no change in optical density is observed, indicating that the LDH and PK are free of enolase activity, it is no longer necessary to carry out the blank determination each time. If, however, some residual enolase activity is present in the auxiliary enzymes, this blank should be determined each day.

If an appreciable blank rate is observed, a blank assay should also be carried out substituting β-mercaptoethanol-EDTA stabilizing solution for hemolyzate in the low S system. This blank should be used in computation of the activity of the low S system.

D. COMMENTS, CALCULATIONS, AND NORMAL VALUES

One mole of PEP is formed for each mole of 2-PGA utilized. This results, ultimately, in the oxidation of 1 mole of NADH to NAD. The equation on page 34 ($\epsilon = 6.22$) is used to calculate the enzyme activity.

The red cells of normal adults contain 5.39 ± 0.83 IU of enolase/g Hb (mean $\pm$ standard deviation) at $37°$. The activity of the low S system is normally $63.1 \pm 9.27\%$ of that in the regular system at $37°$. Factors which permit conversion of results obtained at $25°$ and at $30°$ into values which would have been obtained with the same hemolyzate at $37°$ are presented in Appendix 2.

17

pyruvate kinase (PK)

A. PRINCIPLE

Pyruvate kinase catalyzes the phosphorylation of ADP to ATP by phosphoenolpyruvate (PEP):

$$\text{PEP} + \text{ADP} \xrightarrow[\text{Mg}^{2+},\text{K}^+]{\text{PK}} \text{Pyruvate} + \text{ATP}$$

The rate of formation of pyruvate is measured by linking it to the oxidation of NADH in the lactic dehydrogenase (LDH) reaction:

$$\text{Pyruvate} + \text{NADH} + \text{H}^+ \underset{}{\overset{\text{LDH}}{\rightleftharpoons}} \text{Lactate} + \text{NAD}^+$$

The decrease in optical density which occurs as NADH is oxidized is measured at 340 nm.

Pyruvate kinase is an allosteric enzyme. At low PEP concentrations, its activity is normally stimulated by minute amounts of fructose diphosphate, as the sigmoid substrate-activity curve is converted to a hyperbolic curve. The amount of fructose diphosphate even in very dilute hemolyzates is sufficient to bring about this conversion. Hence, dialyzed hemolyzate is used in the assay.

B. PROCEDURE

The following reagents are added to cuvettes with a critical volume of less than 1 ml. If a large number of assays is to be carried out, multiples of the indicated volumes of tris buffer, KCl, $MgCl_2$, NADH, ADP, and water may be mixed and appropriate quantities of the mixture distributed into cuvettes.

	Blank (μl)	System (μl)	Low S system (μl)	Low S + FDP (μl)
Tris-HCl, 1 M, EDTA, 5 mM, pH 8	100	100	100	100
KCl, 1 M	100	100	100	100
$MgCl_2$, 0.1 M	100	100	100	100
NADH, 2 mM	100	100	100	100
ADP (neut.) 30 mM	—	50	20	20
LDH, 60 U/ml	100	100	100	100
FDP, 10 μM	—	—	—	50
1:20 hemolyzate (dialyzed against β-mercaptoethanol-EDTA stabilizing solution)	20	20	20	20
H_2O	380	330	455	405
Incubate at 37° C for 10 min				
PEP, 50 mM	100	100	5	5

The decrease of optical density of the system, low S system, and low S + FDP system is measured against that of the blank at 340 nm at 37° C.

A recorder expansion giving a full-scale reading of 0.4 OD unit is suitable for normal blood samples. When using a recording spectrophotometer which does not have a provision for arbitrarily adjusting the position of the recorder pen, or when using a nonrecording spectrophotometer, the instrument should be blanked at an optical density value of 0.3, since the optical density of the system will fall.

C. ADDITIONAL BLANKS REQUIRED

A blank assay should be carried out to be certain that the LDH used is free of PK activity. β-Mercaptoethanol-EDTA stabilizing solution is substituted for hemolyzate in both the blank and system mixtures, and the change in optical density of the hemolyzate-free system is measured against that of the hemolyzate-free blank. If no change in optical density is observed, indicating that the LDH is free of PK, it is not necessary to carry out the blank determination each time, as long as the same LDH preparation is being used. If some residual PK activity is present in the auxiliary enzyme, however, this blank should be determined each day.

If an appreciable blank rate is observed, a blank assay should also be carried out substituting β-mercaptoethanol-EDTA stabilizing solution for hemolyzate in the low S and low S + FDP systems. These blanks should be used in computation of the activity of systems.

D. COMMENTS, CALCULATIONS, AND NORMAL VALUES

White cells (WBC) are a rich source of PK, and the activity of this enzyme is not diminished in WBC in red cell PK deficiency. It is therefore particularly important that white cells have been removed, as described on page 10.

The blank rate, if any (see Section C above), should be subtracted from the rate measured in the presence of hemolyzate. One mole of NADH is oxidized for each mole of PEP dephosphorylated. The equation on page 34 ($\epsilon = 6.22$) is used to calculate the enzyme activity.

The red cells of normal adults contain 15.0 ± 1.99 IU of PK/g Hb (mean $\pm$ standard deviation) at $37°$. The activity of the low S system is normally $14.9 \pm 3.71\%$ and the activity of the low S + FDP system is normally $43.5 \pm 2.46\%$ of that in the regular system at $37°$. Factors which permit conversion of results obtained at $25°$ and at $30°$ into values which would have been obtained with the same hemolyzate at $37°$ are presented in Appendix 2.

PK deficiency is probably the most common single known cause of nonspherocytic congenital hemolytic anemia. Mutations which result in the formation of an enzyme with near-normal activity at high substrate concentration but diminished activity with low substrate concentration are known to occur. Mutants in which the principal abnormality is an inadequate response to the allosteric effector, fructose diphosphate, also occur. Young red cells normally have increased levels of pyruvate kinase, and for these reasons, the enzymatic diagnosis of pyruvate kinase deficiency may be particularly difficult.

Most pyruvate kinase variants are unstable to heating. We have found a heat stability test [1] to be a particularly helpful adjunct in the diagnosis of pyruvate kinase deficiency. To carry out this procedure, nine parts of the 1:20 hemolysate are mixed with one part of the 1 M tris HCl-5 mM EDTA, pH 8.0, buffer. A control sample is removed before heating, and additional samples are removed after heating for 20, 40, and 60 minutes at 56° C. Each sample is centrifuged at 7000 g for 10 min. to remove the precipitate which has formed, and PK assays are carried out in the usual manner. In fresh samples from normal donors 68.3 ± 4.8% of the enzyme remains at 20 minutes, 49.9 ± 3.7% at the end of 40 minutes, and 36.2 ± 1.8% at 60 minutes. It is not necessary to carry out assays in the low S system or low S + FDP system for this purpose.

REFERENCE

1. Blume, K. G., Arnold, H., Löhr, G. W., and Beutler, E.: Additional diagnostic procedures for the detection of abnormal red cell pyruvate kinase. Clin. Chim. Acta 43:443–446, 1973.

18

lactate dehydrogenase (LDH)

A. PRINCIPLE

Lactate dehydrogenase catalyzes the reduction of pyruvate to lactate by NADH:

$$\text{Pyruvate} + \text{NADH} + \text{H}^+ \underset{}{\overset{\text{LDH}}{\rightleftharpoons}} \text{Lactate} + \text{NAD}^+$$

In the assay the oxidation of NADH is followed spectrophotometrically at 340 nm.

B. PROCEDURE

The following reagents are added to cuvettes with a critical volume of less than 1 ml. If a large number of assays is to be carried out, multiples of the indicated volumes of tris buffer, NADH, and water may be mixed and appropriate quantities of the mixture distributed into cuvettes.

	Blank (μl)	System (μl)
Tris-HCl, 1 M, EDTA, 5 mM, pH 8.0	100	100
NADH, 2 mM	100	100
1:200 hemolyzate	20	20
H$_2$O	780	680

Incubate at 37° C for 10 min

Sodium pyruvate, 10 mM	—	100

The decrease of optical density of the system is measured against that of the blank at 340 nm at 37° C.

A recorder expansion giving a full-scale reading of 1.0 OD unit is suitable for normal blood samples. When using a recording spectrophotometer which does not have a provision for arbitrarily adjusting the position of the recorder pen, or when using a nonrecording spectrophotometer, the instrument should be blanked at an optical density value of 0.6, since the optical density of the system will fall.

C. ADDITIONAL BLANKS REQUIRED

None.

D. COMMENTS, CALCULATIONS, AND NORMAL VALUES

One mole of NADH is oxidized for each mole of pyruvate reduced. The equation on page 34 ($\epsilon = 6.22$) is used to calculate the enzyme activity. The red cells of normal adults contain 200 ± 26.5 IU of LDH/g Hb (mean $\pm$ standard deviation) at 37°. Factors which permit conversion of results obtained at 25° and at 30° into values which would have been obtained with the same hemolyzate at 37° are presented in Appendix 2.

part IV

red cell
hexose monophosphate shunt
and other enzymes

19

glucose-6-phosphate dehydrogenase (G-6-PD) and 6-phosphogluconate dehydrogenase (6-PGD)

A. PRINCIPLE

Glucose-6-phosphate dehydrogenase catalyzes the oxidation of glucose-6-P to 6-phosphogluconolactone which quickly hydrolyzes spontaneously to 6-phosphogluconate (6-PGA):

$$\text{Glucose-6-P} + \text{NADP}^+ \xrightarrow{\text{G-6-PD}} \text{6-PGA} + \text{NADPH} + \text{H}^+$$

6-Phosphogluconate dehydrogenase catalyzes the oxidation of 6-PGA to ribulose-5-phosphate and CO_2:

$$\text{6-PGA} + \text{NADP}^+ \xrightarrow{\text{6-PGD}} \text{Ribulose-5-P} + CO_2 + \text{NADPH} + \text{H}^+$$

The most commonly used assays for G-6-PD activity measure the rate of reduction of NADP to NADPH when a hemolyzate is incubated with glucose-6-P. However, a portion of the product formed, 6-phosphogluconate, is largely oxidized further in the 6-PGD reaction, so that more than 1 (nearly 2) moles of NADP are reduced for each mole of glucose-6-P oxidized. This type of assay was recommended by a WHO study group [1] for reasons discussed below. The assay results obtained can be corrected for the oxidation of 6-PGA by adding a saturating quantity of 6-PGA along with glucose-6-P to the assay system, and subtracting, from the rate of NADP reduction observed, the rate of reduction in a system to which 6-PGA alone has been added. This type of assay procedure was first described by Glock and McLean [2] and is referred to as a Glock and McLean assay. The four-cuvette assay system described below permits at the same time: (1) the calculation of G-6-PD activity in the standard way, (2) the G-6-PD activity as corrected for 6-PGD activity, and (3) the 6-PGD activity.

B. PROCEDURE

The following reagents are added to cuvettes with a critical volume of less than 1 ml. If a large number of assays is to be carried out, multiples of the indicated volumes of tris buffer, $MgCl_2$, NADP, and water may be mixed and appropriate quantities of the mixture distributed into cuvettes.

	Cuvette			
	1 (μl)	2 (μl)	3 (μl)	4 (μl)
Tris-HCl, 1 M, EDTA, 5 mM, pH 8.0	100	100	100	100
$MgCl_2$, 0.1 M	100	100	100	100
NADP, 2 mM	100	100	100	100
1:20 hemolyzate	20	20	20	20
H_2O	680	580	580	480

Incubate at 37° C for 10 min

G-6-P, 6 mM	—	100	—	100
6-PGA, 6 mM	—	—	100	100

	Cuvette			
Low S system	1 (μl)	2 (μl)	3 (μl)	4 (μl)
Tris-HCl, 1 M, EDTA, 5 mM, pH 8.0	100	100	100	100
$MgCl_2$, 0.1 M	100	100	100	100
NADP, 2 mM	100	100	100	100
1:20 hemolyzate	20	20	20	20
H_2O	680	670	675	665

Incubate at 37° C for 10 min

G-6-P, 6 mM	—	10	—	10
6-PGA, 6 mM	—	—	5	5

The increase of optical density of cuvettes 2, 3, and 4 is measured against that of cuvette 1 at 340 nm at 37°C. A recorder expansion giving a full-scale reading of 1.0 OD unit is suitable for normal blood samples.

C. ADDITIONAL BLANKS REQUIRED

None.

D. COMMENTS, CALCULATIONS, AND NORMAL VALUES

The assay values obtained using only G-6-P as a substrate resemble those obtained in the standard WHO recommended assay for G-6-PD. They differ in that the temperature at which the assay is carried out is 37° rather than

25°. This results not only in a more rapid rate because of the higher temperature but also in a somewhat lower cuvette pH than is present in the WHO assay. For this reason, the assay is referred to as the WHO-37°.

The result obtained by subtracting the 6-PGD activity from the activity obtained with both substrates (Glock and McLean) is normally a more accurate expression of G-6-PD activity. However, in the case of samples with very low G-6-PD activity, as found in hereditary G-6-PD deficiency, this requires subtracting a large, experimentally obtained value from a slightly larger one. This results in great magnification of errors, and the method is therefore unsuitable for samples with very low G-6-PD activity. The only other differences between the standard WHO assay and the assay procedure given above are that EDTA is included in the tris buffer, and the preparation of the hemolyzate is different. Neither of these changes materially affects the results obtained on normal hemolyzates. The absence of NADP, included in the stabilizing solution used in the standard WHO method, does not adversely affect the stability of normal hemolyzate G-6-PD kept at 0° for several hours. This omission may, however, result in loss of activity from some abnormal G-6-PD variants. Thus with most samples one may duplicate the values which would be obtained in the standard WHO recommended procedure, by lowering the temperature at which the assay is carried out from 37° to 25°, or by making the appropriate correction using the conversion factor in Appendix 2. In characterizing mutant enzymes, however, use of the standard procedure [1] is recommended.

Although nearly 2 moles of NADP are reduced for each mole of glucose-6-P oxidized, calculations of WHO-37° assay values are based on the assumption that only 1 mole of NADP is reduced. This illogical assumption is made to conform with international usage. The enzyme activity for G-6-PD-WHO-37° is therefore calculated from the rate in cuvette 2 using the equation on page 34 ($\epsilon = 6.22$). In the 6-phosphogluconate dehydrogenase reaction 1 mole of NADP is reduced for each mole of 6-PGA oxidized. The enzyme activity is therefore also calculated with $\epsilon = 6.22$.

The G-6-PD Glock and McLean activity is obtained by subtracting the rate obtained in cuvette 3 from the rate obtained in cuvette 4 and calculating the activity according to the equation on page 34 ($\epsilon = 6.22$).

The red cells of normal adults contain 8.34 ± 1.59 IU of G-6-PD/g Hb (mean $\pm$ standard deviation) at 37° (Glock and McLean method). The activity of the Glock and McLean low S system is normally $67.1 \pm 6.53\%$ of that in the regular system at 37°. The red cells of normal adults contain 12.1 ± 2.09 IU of G-6-PD/g Hb (mean $\pm$ standard deviation) at 37° (WHO method). The activity of the WHO low S system is normally $62.2 \pm 4.65\%$ of that in the regular system at 37°. The red cells of normal adults contain 8.78 ± 0.78 IU of 6-PGD/g Hb (mean $\pm$ standard deviation) at 37°. The activity of the low S system is normally $62.4 \pm 4.21\%$ of that in the regular system at 37°. Factors which permit conversion of results obtained at 25° and at 30° into values which would have been obtained with the same hemolyzate at 37° are presented in Appendix 2.

G-6-PD deficiency is the most common enzymatic deficiency of the red cell. It results in sensitivity to drug-induced hemolytic anemia, favism, neonatal icterus, and in some cases nonspherocytic congenital hemolytic anemia.

Severe 6-PGD deficiency is extremely rare, and its clinical effects have not been clearly demarcated.

REFERENCES

1. Report of a WHO Scientific Group: Standardization of procedures for the study of glucose-6-phosphate dehydrogenase. WHO Techn. Rep. Ser. 366, 1967.
2. Glock, G. E., and McLean, P.: Further studies on the properties and assay of glucose-6-phosphate dehydrogenase and 6-phosphogluconate dehydrogenase of rat liver. Biochem. J. 55:400-408, 1953.

20

glutathione reductase (GR)

A. PRINCIPLE

Glutathione reductase catalyzes the reduction of oxidized glutathione (GSSG) by NADPH or NADH to reduced glutathione (GSH):

$$\text{NADPH (NADH)} + \text{H}^+ + \text{GSSG} \xrightarrow{\text{GR}} \text{NADP}^+ \text{ (NAD}^+) + 2\text{GSH}$$

The activity of the enzyme is measured by following the oxidation of NADPH spectrophotometrically at 340 nm. GR is a flavin enzyme, and it has been found that it is not fully activated by FAD in normal hemolyzates. Complete activation of apoenzyme requires the preincubation of enzyme with FAD. This must be done before GSSG or NADPH is added to the reaction system, since these seem to interfere with activation of the enzyme by FAD.

B. PROCEDURE

The following reagents are added to cuvettes with a critical volume of less than 1 ml

	Blank (μl)	System (μl)	Blank (μl)	System (μl)
Tris-HCl, 1 M, EDTA, 5 Mm, pH 8.0	50	50	50	50
1 : 20 hemolyzate	10	10	10	10
H₂O	890	790	790	690
FAD, 10 μM	—	—	100	100
Incubate at 37° for 10 min				
GSSG, 0.033 M (neut.)	—	100	—	100
Incubate at 37° for 10 min				
NADPH (spectrophotometrically standardized*), 2 mM	50	50	50	50

The decrease of optical density at 340 nm, 37° C, of the system without FAD is measured against that of the blank without FAD. The system with FAD is measured in the same way against the blank with FAD.

A recorder expansion giving a full-scale reading of 0.2 OD unit is suitable for normal blood samples. When using a recording spectrophotometer which does not have a provision for arbitrarily adjusting the position of the recorder pen, or when using a nonrecording spectrophotometer, the instrument should be blanked at an optical density value of 0.15, since the optical density of the sytem will fall.

C. ADDITIONAL BLANKS REQUIRED

None.

D. COMMENTS, CALCULATIONS, AND NORMAL VALUES

The enzyme frequently becomes somewhat more active as the assay proceeds. Thus, an increasing rate is often observed for the first few minutes.

One mole of NADPH is oxidized for each mole of GSSG reduced. The equation on page 34 ($\epsilon = 6.22$) is used to calculate the enzyme activity. The red cells of normal adults contain 7.18 ± 1.09 IU of GR/g Hb (mean ± standard deviation) at 37° without FAD, and 10.4 ± 1.50 IU of GR/g Hb at 37° with added FAD. Factors which permit conversion of results obtained at 25° and at 30° into values which would have been obtained with the same hemolyzate at 37° are presented in Appendix 2.

* see page 21.

The extent of activation of GR by FAD is a guide to the adequacy of riboflavin nutrition. Low GR activity may occur on a hereditary basis, but most frequently can be corrected in vitro with FAD and in vivo by the administration of 5 mg of riboflavin a day for a few days [1].

REFERENCE

1. Beutler, E.: Effect of flavin compounds on glutathione reductase activity: in vivo and in vitro studies. J. Clin. Invest. 48:1957–1966, 1969.

21

glutathione peroxidase (GSH-Px)

A. PRINCIPLE

Glutathione peroxidase catalyzes the oxidation of GSH to GSSG by hydrogen peroxide:

$$2\text{GSH} + \text{R—O—O—H} \xrightarrow{\text{GSH-Px}} \text{GSSG} + \text{H}_2\text{O} + \text{R—OH}$$

where R—O—O—H is a peroxide. t-Butyl hydroperoxide is the most suitable substrate for assay of the enzyme. The rate of formation of GSSG is measured by means of the glutathione reductase reaction:

$$\text{GSSG} + \text{NADPH} + \text{H}^+ \xrightarrow{\text{GR}} 2\text{GSH} + \text{NADP}^+$$

The oxidation of NADPH is followed at 340 nm.

B. PROCEDURE

The following reagents are added to cuvettes with a critical volume of less than 1 ml. If a large number of assays is to be carried out, multiples of the indicated volumes of buffer, GSH, and water may be mixed and appropriate quantities of the mixture distributed into cuvettes.

	Blank	System
Tris-HCl, 1 M, EDTA, 5 mM, pH 8.0	100	100
GSH, 0.1 M	20	20
Gutathione reductase, 10 U/ml	100	100
NADPH, 2 mM	100	100
1:20 hemolyzate	10	10
H$_2$O	670	660

Preincubate at 37° C for 10 min

| t-Butyl hydroperoxide, 7 mM* | — | 10 |

* Approximately 1:1000 dilution.

The decrease of OD of the system is measured against that of the blank at 340 nm.

A recorder expansion giving a full-scale reading of 1.0 OD unit is suitable for normal blood samples. When using a spectrophotometer which does not have a provision for arbitrarily adjusting the position of the recorder pen, or when using a nonrecording spectrophotometer, the instrument should be blanked at an optical density value of 0.6, since the optical density of the system will fall.

C. ADDITIONAL BLANKS REQUIRED

A blank assay substituting β-mercaptoethanol-EDTA stabilizing solution for hemolyzate should be carried out in order to correct for nonenzymatic oxidation of GSH and NADPH by t-butyl hydroperoxide. This rate will be approximately 0.015 to 0.02 OD unit/min.

D. COMMENTS, CALCULATIONS, AND NORMAL VALUES

The rate of the reaction is strongly dependent on the concentration of t-butyl hydroperoxide. The 7 mM t-butyl hydroperoxide should be prepared daily. The concentration of pure tertiary butyl hydroperoxide, as purchased, is calculated to be 10 M. Thus, a 1:1000 dilution should provide a tertiary butyl hydroperoxide solution of 10 mM concentration. However, titration of this material with thiosulfate according to the method described by Kokatnur and Jelling [1], showed that the commercial preparation was only approximately 70% hydroperoxide. Since the purity of various lots of this material may differ, it is desirable to titrate each lot individually to obtain reproducible results from lot to lot. t-Butyl hydroperoxide may be obtained from SciChem Co., Frese Division, 1430 Grande Vista Ave., Los Angeles, 90023 and Koch-Light, Research Products International Corp., 2692 Delta Lane, Elk Grove Village, Ill. 60007. GSH solutions must be freshly prepared, but even so they always contain some contaminating GSSG. This results in some oxidation of NADPH. If too much GSSG is present in the GSH, no change in absorbance will be noted after addition of peroxide, or a small change will occur and then cease. In this case, a more pure GSH solution must be used or additional NADPH added to the assay system.

The blank rate (see Section C above) should be subtracted from the rate measured in the presence of hemolyzate. One mole of NADPH is oxidized for each mole of t-butyl hydroperoxide reduced. The equation on page 34 ($\epsilon = 6.22$) is used to calculate the activity.

The red cells of normal U.S.-Northern European and U.S. African adults contain 30.8 ± 4.73 IU of GSH-Px/g Hb (mean $\pm$ standard deviation) at 37° when the blood is collected in EDTA. Blood collected in ACD gives similar values (normal = 32.07 ± 3.84), but significantly higher GSH-Px activity is observed with heparinized blood (34.2 ± 4.77). There are considerable ethnic differences in GSH-Px activity. Among normal U.S.-Jewish and Israeli-Jewish adults 23.77 ± 6.97 IU of GSH-Px/gm Hb were found, apparently due to the existence of a low GSH-Px polymorphism. Among normal U.S.-Oriental adults 24.52 ± 2.74 IU of GSH-Px/gm Hb were found. Factors which permit conversion of results obtained at 25° and at 30° into values which would have been obtained with the same hemolyzate at 37° are presented in Appendix 2. Glutathione peroxidase deficiency resulting in nonspherocytic congenital hemolytic anemia has been reported in the literature. It has also been suggested that low glutathione peroxidase levels in the red cells of newborn infants may contribute to hemolytic disease of the newborn.

REFERENCE

1. Kokatnur, V. R., and Jelling, M.: Iodometric determination of peroxygen in organic compounds. Amer. Chem. Soc. 63:1432–1433, 1941.

22

NADPH diaphorase

A. PRINCIPLE

This enzyme catalyzes the transfer of hydrogen from NADPH to methylene blue (MeBl), reducing it to leukomethylene blue (LeukMeBl):

$$\text{NADPH} + \text{H}^+ + \text{MeBl} \xrightarrow{\text{NADPH diaphorase}} \text{NADP}^+ + \text{LeukMeBl}$$

The rate of reaction can be followed by measuring the oxidation of NADPH at 340 nm.

B. PROCEDURE (based on Huennekens et al. [1])

The following reagents are added to cuvettes with a critical volume of less than 1 ml.

	Blank (μl)	System (μl)
Tris-HCl, 1 M, EDTA, 5 mM, pH 8.0	100	100
NADPH, 2 mM	—	20
1:20 hemolyzate	50	50
H$_2$O	840	820
Incubate at 37° for 10 min		
Methylene blue, 0.8 mM	10	10

The decrease in the optical density of the system is measured against that of the blank at 340 nm at 37°C.

A recorder expansion giving a full-scale reading of 0.4 OD unit is suitable for normal blood samples. When using a recording spectrophotometer which does not have a provision for arbitrarily adjusting the position of the recorder pen, or when using a nonrecording spectrophotometer, the instrument should be blanked at an optical density value of 0.3, since the optical density of the system will fall.

74

C. ADDITIONAL BLANKS REQUIRED

To correct for the slow nonenzymatic oxidation of NADPH by methylene blue, water is substituted for hemolyzate in both the blank and system mixtures and the change in optical density of the system is measured against that of the blank. This rate will be approximately 0.023 OD unit per minute.

D. COMMENTS, CALCULATIONS, AND NORMAL VALUES

The minute quantity of MeBl in the reaction mixture suffices because the LeukMeBl is rapidly reoxidized by molecular oxygen. With the aging of methylene blue preparations, the blank rate sometimes becomes more rapid. When this occurs, a fresh methylene blue solution should be prepared.

The blank rate (see Section C above) should be subtracted from the rate measured in the presence of hemolyzate. The equation on page 34 ($\epsilon = 6.22$) is used in making the calculations. The red cells of normal adults contain 2.26 ± 0.16 IU of NADPH diaphorase/g Hb (mean $\pm$ standard deviation) at 37°.

One case of deficiency of this enzyme has been reported. Absence of this enzyme will result in false positive results for some of the common screening tests for glucose-6-phosphate dehydrogenase deficiency, such as the methemoglobin reduction test or the brilliant cresyl blue reduction test. However, the enzyme deficiency produces no clinical effects.

REFERENCES

1. Huennekens, F. M., Caffrey, R. W., Basford, R. E., and Gabrio, B. W.: Erythrocyte metabolism. IV. Isolation and properties of methemoglobin reductase. J. Biol. Chem. 227:261–272, 1957.

23

NADH methemoglobin reductase (NADH diaphorase)

A. PRINCIPLE

NADH diaphorase normally catalyzes the reduction of methemoglobin by NADH:

$$Hb^{3+} + H^+ + NADH \xrightarrow{\text{NADH diaphorase}} Hb^{2+} + NAD^+$$

However, this reaction proceeds only very slowly, and is difficult to follow spectrophotometrically. It has been found, however, that a methemoglobin-ferrocyanide complex serves as an excellent substrate for the enzyme, and that its activity can be followed spectrophotometrically by measuring the reduction of the methemoglobin-ferrocyanide complex at 575 nm.

B. PROCEDURE (based on Hegesh et al. [1])

A hemoglobin determination is carried out on the whole blood sample to be assayed.

The following reagents are mixed in a centrifuge tube in the quantities specified or multiples thereof.

	ml
EDTA, 0.27 M (neut.)	0.010
Sodium citrate, 0.05 M, pH 4.7	0.50
$K_3Fe(CN)_6$, 0.5 mM	1.50
Hb substrate*	1.00
H_2O	1.74

* The Hb substrate is hemoglobin which has been largely freed of NADH diaphorase activity. Centrifuge heparinized or citrated blood. Discard plasma and wash the red cells two times in 10 volumes of saline solution. To each 10 ml of washed, packed cells, add 60 ml of water and 1.6 g of dry DEAE-cellulose (e.g., Whatman DE-11). Let stand

Add 5 μl of whole blood to the 4.75-ml reaction mixture, and centrifuge for 10 min at 9500 g at 4° C. Using the supernatant (blood-assay mixture), set up the following system in cuvettes with a critical volume of less than 1 ml:

	Blank (μl)	System (μl)
Blood-assay mixture	950	950
H$_2$O	50	—
Incubate at 37° C for 10 min		
NADH, 2 mM	—	50

The increase in optical density of the system is read against that of the blank at 575 nm at 37° C.

A recorder expansion giving a full-scale reading of 1.0 OD unit is suitable for normal blood samples.

C. ADDITIONAL BLANKS REQUIRED

It is necessary daily to determine the extent of contamination of the hemoglobin substrate with NADH diaphorase. In order to do this, a blank mixture is prepared which is identical in composition to the blood-assay mixture, except that the volume of hemoglobin substrate is increased to 1.05 ml, and only 1.69 ml of water are added. No blood is added to this mixture, and an assay is carried out in the regular manner.

D. COMMENTS, CALCULATIONS, AND NORMAL VALUES

Although this method is more cumbersome to carry out than most of the other red cell enzyme assays described, it is currently the best procedure available for the assay of NADH diaphorase activity. The ratio of potassium ferricyanide to hemoglobin substrate is fairly critical. If too much ferricyanide is present, the onset of the reaction will be delayed and it will be slowed.

The blank rate, if any (see Section C above), should be subtracted from the rate measured in the presence of hemolyzate. Calculation of enzyme activ-

---

10 min, mixing occasionally, and filter. Repeat two times by adding to the filtrate a new portion of 1.6 g DEAE-cellulose per 10 ml of cells. Determine the hemoglobin concentration of the clear supernatant in Drabkin's solution, and adjust the final concentration to 1.224 g %. Run blank assay employing 1.05 ml of hemoglobin substrate and omitting the addition of the blood sample. Very little NADH diaphorase activity (ΔOD < 0.003/min) should remain in the solution. If activity is greater, repeat treatment with DEAE as often as required. The purified hemoglobin substrate is stable for 2 weeks if kept refrigerated.

ity is based on an optical density difference between ferro- and ferrihemo-globin of 42 for a 1 mM solution. The activity (A) in micromoles of Hb reduced in the 1-ml system is then obtained as follows:

$$A = \frac{(\Delta OD_R - \Delta OD_B)}{420}$$

where ΔOD is the change in optical density at 575 nm per 10 min, in the complete system, and ΔOD_B is the change in the blank system without blood (see Section C, above). The activity (E) in international units per gram of Hb is

$$E = \frac{A}{Hb_F}$$

where Hb_F is the number of grams of hemoglobin in the final assay system.

$$Hb_F = Hb \times \frac{1}{100} \times \frac{0.005}{4.755} \times 0.95 = Hb \times 10^{-5}$$

where Hb is the hemoglobin concentration of the blood sample in gram per-cent; 1/100 converts the Hb concentration to grams per milliliter, while the other two factors take into account the dilution of the sample. Therefore,

$$E = \frac{\Delta OD_R - \Delta OD_B}{Hb} \times 2.38 \times 10^2$$

or if the volume of blood added (V_H) is varied:

$$E = \frac{\Delta OD_R - \Delta OD_B}{Hb} \times \frac{1.19}{V_H}$$

The red cells of normal adults contain 3.40 ± 0.50 IU of NADH-methemo-globin reductase/g Hb (mean ± standard deviation) at 37°. Factors which permit conversion of results obtained at 25° and at 30° into values which would have been obtained with the same hemolyzate at 37° are presented in Appendix 2.

REFERENCE

1. Hegesh, E., Calmanovici, N., and Avron, M.: New method for determining ferrihe-moglobin reductase (NADH-methemoglobin reductase) in erythrocytes. J. Lab. Clin. Med. 72:339–344, 1968.

24

phosphoglucomutase (PGM)

A. PRINCIPLE

Phosphoglucomutase catalyzes the interconversion of glucose-1-P to glucose-6-P:

$$\text{Glucose-1-P} \underset{\text{G-1-6-diP}}{\overset{\text{PGM}}{\rightleftharpoons}} \text{Glucose-6-P}$$

In this assay, glucose-1-P serves as substrate, and the formation of glucose-6-P is measured by linking its further oxidation to 6 PGA with reduction of NADP through the glucose-6-P dehydrogenase reaction:

$$\text{Glucose-6-P} + \text{NADP}^+ \xrightarrow{\text{G6PD}} \text{6-Phosphogluconate} + \text{NADPH} + \text{H}^+$$

Some of the 6-phosphogluconate which is formed is then oxidized further by the phosphogluconic dehydrogenase, present in almost all hemolyzates, reducing more NADP to NADPH:

$$\text{6-Phosphogluconate} + \text{NADP}^+ \xrightarrow{\text{6-PGD}} \text{Ribulose-5-P} + \text{NADPH} + \text{H}^+$$

The reduction of NADP is followed at 340 nm.

B. PROCEDURE

The following reagents are added to cuvettes with a critical volume of less than 1 ml. If a large number of assays is to be carried out, multiples of the indicated volumes of tris buffer, $MgCl_2$, NADP, glucose-1,6-diP, and water may be mixed and appropriate quantities of the mixture distributed into cuvettes.

	Blank (μl)	System (μl)
Tris-HCl, 1 M, EDTA, 5 mM, pH 8.0	100	100
MgCl$_2$, 0.1 M	50	50
NADP, 2 mM	100	100
1:20 hemolyzate	50	50
G-6-PD, 10 U/ml (diluted in β-mercaptoethanol-EDTA stabilizing solution)	20	20
G-1,6-diP, 0.7 mM	200	200
H$_2$O	480	430

Incubate at 37° for 10 min

| Glucose-1-P, 0.05 M | — | 50 |

The increase in the optical density of the system is measured against that of the blank at 340 nm at 37° C. A recorder expansion giving a full-scale reading of 1.0 OD unit is suitable for normal blood samples.

C. ADDITIONAL BLANKS REQUIRED

A blank assay should be carried out to be certain that the G-6-PD used is free of PGM activity. β-Mercaptoethanol-EDTA stabilizing solution is substituted for hemolyzate in both the blank and system mixtures, and the change in optical density of the hemolyzate-free system is measured against that of the hemolyzate-free blank.

If no change in optical density is observed, indicating that the G-6-PD is free of PGM activity, it is not necessary to carry out the blank determination each time as long as the same G-6-PD preparation is being used. If some residual activity is present in the auxiliary enzyme, however, this blank should be determined each day.

D. COMMENTS, CALCULATIONS, AND NORMAL VALUES

Glucose-1,6-diP, an essential cofactor for the reaction, is very expensive and is sometimes difficult to obtain commercially. If it cannot be obtained, a boiled extract of fresh red cells is a fairly satisfactory substitute, since it is quite rich in glucose-1-6-diP.

The blank rate, if any (see Section C above), should be subtracted from the rate measured in the presence of hemolyzate. Accurate computation of the amount of glucose-1-phosphate converted to glucose-6-phosphate is not possible, since conversion of 6-phosphogluconate to ribulose-5-P is incomplete. Thus, between 1 and 2 moles of NADP are reduced to NADPH. Actual measurement of the amount of 6-phosphogluconate which accumulates, indicates that most of the 6-phosphogluconate is, in fact, oxidized to reduce NADP. The equation on page 34 ($\epsilon = 12.44$) is used to calculate the results.

PGM plays an important role in other tissues in glycogen metabolism and in the conversion of galactose to glucose. The role of this enzyme in red cell metabolism is not clearly established. One case of partial PGM deficiency has been reported.

The red cells of normal adults contain 5.50 ± 0.62 IU of PGM/g Hb (mean $\pm$ standard deviation) at $37°$. Factors which permit conversion of results obtained at $25°$ and at $30°$ into values which would have been obtained with the same hemolyzate at $37°$ are presented in Appendix 2.

25

glutamate-oxaloacetate transaminase (GOT)

A. PRINCIPLE

This enzyme catalyzes the reaction

$$\text{L-Aspartate} + \alpha\text{-Ketoglutarate} \xrightarrow{\text{GOT}} \text{Oxaloacetate} + \text{L-Glutamate}$$

The oxaloacetate formed is measured in the malic dehydrogenase (MDH) reaction:

$$\text{Oxaloacetate} + H^+ + \text{NADH} \xrightarrow{\text{MDH}} \text{Malate} + \text{NAD}^+$$

As with other transaminases, pyridoxal phosphate serves as a cofactor for the activity of GOT. The increase of enzyme activity after the addition of this compound serves as an index of the degree of saturation of apoenzyme with cofactor.

B. PROCEDURE

The following reagents are added to cuvettes with a critical volume of less than 1 ml. If a large number of assays is to be carried out, multiples of the

indicated volumes of tris buffer, NADH, L-aspartate, and water may be mixed
and appropriate quantities of the mixture distributed into cuvettes.

	Without pyridoxal phosphate		With pyridoxal phosphate	
	Blank (μl)	System (μl)	Blank (μl)	System (μl)
Tris-HCl, 1 M, EDTA, 5 mM, pH 8.0	100	100	100	100
NADH, 2 mM	100	100	100	100
L-Aspartate, pH 8, 0.1 M	100	100	100	100
Malic dehydrogenase, 1 U/ml	10	10	10	10
Pyridoxal 5'-phosphate, 0.4 mM	0	0	50	50
H$_2$O	670	570	620	520
1:20 hemolyzate	20	20	20	20
Incubate at 37° C for 10 min				
Na α-ketoglutaric acid, 0.1 M	0	100	0	100

The decrease in the optical density of each system is measured against that
of each blank at 340 nm at 37° C.

A recorder expansion giving a full-scale reading of 0.4 OD unit is suitable
for normal blood samples. When using a recording spectrophotometer which
does not have a provision for arbitrarily adjusting the position of the recorder
pen, or when using a nonrecording spectrophotometer, the instrument should
be blanked at an optical density value of 0.3, since the optical density of
the system will fall.

C. ADDITIONAL BLANKS REQUIRED

A blank assay should be carried out to be certain that the MDH used is
free of GOT activity. β-Mercaptoethanol-EDTA stabilizing solution is substituted for hemolyzate in both the blank and system mixtures, and the change
in optical density of the hemolyzate-free system is measured against that of
the hemolyzate-free blank. If no change in optical density is observed, indicating that the MDH is free of GOT, it is not necessary to carry out the blank
determination each time as long as the same MDH preparation is being used.
If some residual GOT activity is present in the auxiliary enzyme, however,
this blank should be determined each day.

D. COMMENTS, CALCULATIONS, AND NORMAL VALUES

The maximum rate may not be reached for 10 to 15 min. The blank rate,
if any (see Section C above), should be subtracted from the maximum rate
measured in the presence of hemolyzate. One mole of NADH is oxidized

for each mole of glutamate transaminated. The equation on page 34 ($\epsilon =$ 6.22) is used.

The red cells of normal adults contain 3.02 ± 0.67 IU of GOT/g Hb (mean $\pm$ standard deviation) at 37°, and 5.04 ± 0.90 IU of GOT/g Hb when stimulated with pyridoxal phosphate. Factors which permit conversion of results obtained at 25° and at 30° into values which would have been obtained with the same hemolyzate at 37° are presented in Appendix 2.

The enzyme is commonly used as indicator of mean red cell age, since its activity is considerably higher in young red cells than in older cells. The degree of activation of GOT by pyridoxal phosphate may give some indication of the state of pyridoxine nutrition.

26

adenylate kinase (AK)

A. PRINCIPLE

Adenylate kinase (myokinase) catalyzes the dismutation of ADP into AMP and ATP:

$$2\text{ADP} \underset{\text{Mg}^{2+}}{\overset{\text{AK}}{\rightleftharpoons}} \text{ATP} + \text{AMP}$$

In this assay, the reverse reaction is measured, and the formation of ADP is followed in the pyruvate kinase (PK) and lactate dehydrogenase (LDH) reactions:

$$\text{ADP} + \text{PEP} \overset{\text{PK}}{\rightarrow} \text{Pyruvate} + \text{ATP}$$
$$\text{Pyruvate} + \text{H}^+ + \text{NADH} \xrightarrow{\text{LDH}} \text{Lactate} + \text{NAD}^+$$

The oxidation of NADH is followed at 340 nm.

B. PROCEDURE (based on Levin and Beutler [1])

The following reagents are added to cuvettes with a critical volume of less than 1 ml. If a large number of assays is to be carried out, multiples of the indicated volumes of tris buffer, ATP, NADH, $MgCl_2$, PEP, KCl, and water may be mixed and appropriate quantities of the mixture distributed into cuvettes.

	Blank (μl)	System (μl)	Low S system (μl)
Tris-HCl, 1 M, EDTA, 5 mM, pH 8.0	100	100	100
ATP (neut.), 0.02 M	50	50	10
NADH, 2 mM	100	100	100
PEP, 0.05 M	50	50	50
KCl, 1 M	20	20	20
LDH, 60 U/ml	50	50	50
$MgCl_2$, 0.1 M	50	50	50
PK (Sigma, type II), 50 U/m^l	20	20	20
1:200 hemolyzate	5	5	5
H_2O	555	505	585

Incubate at 37° C for 10 min

AMP (neut.), 0.02 M	—	50	10

The decrease of optical density of the system and low S system is measured against that of the blank at 340 nm at 37° C for 10 to 20 min.

A recorder expansion giving a full-scale reading of 0.4 OD units is suitable for normal blood samples. When using a recording spectrophotometer which does not have a provision for arbitrarily adjusting the position of the recorder pen, or when using a nonrecording spectrophotometer, the instrument should be blanked at an optical density value of 0.3, since the optical density of the system will fall.

C. ADDITIONAL BLANKS REQUIRED

A blank assay should be carried out to be certain that the PK used is free of AK activity. β-Mercaptoethanol-EDTA stabilizing solution is substituted for hemolyzate in both the blank and system mixtures, and the change in optical density of the hemolyzate-free system is measured against that of the hemolyzate-free blank. If no change in optical density is observed, indicating that the PK is free of AK, it is not necessary to carry out the blank determination each time as long as the same PK preparation is being used. If some residual AK activity is present in auxiliary enzyme, however, this blank should be determined each day.

If an appreciable blank rate is observed, a blank assay should also be carried out substituting β-mercaptoethanol-EDTA stabilizing solution for hemolyzate in the low S system. This blank should be used in computation of the activity of the low S system.

D. COMMENTS, CALCULATIONS, AND NORMAL VALUES

The blank rate, if any (see Section C above), should be subtracted from the rate measured in the presence of hemolyzate. Two moles of NADH are oxidized for each mole of ATP reacting with AMP. The equation on page 34 ($\epsilon = 12.44$) is used.

The red cells of normal adults contain 258 ± 29.3 IU of AK/g Hb (mean $\pm$ standard deviation) at 37°. The activity of the low S system is normally $38.0 \pm 3.63\%$ of that in the regular system at 37°. Factors which permit conversion of results obtained at 25° and at 30° into values which would have been obtained with the same hemolyzate at 37° are presented in Appendix 2.

AK deficiency resulting in mild nonspherocytic congenital hemolytic anemia has been reported to occur.

REFERENCE

1. Levin, E., and Beutler, E.: Human erythrocyte adenylate kinase. Haematol. Hungarica 1:19–25, 1967.

27

adenosine deaminase (AD)

A. PRINCIPLE

Adenosine deaminase catalyzes the deamination of adenosine to inosine.

$$\text{Adenosine} + \text{H}_2\text{O} \xrightarrow{\text{AD}} \text{Inosine} + \text{NH}_3$$

The absorption maximum of adenosine is at 265 nm. The activity of the enzyme is measured by following decrease of optical density at 265 nm.

B. PROCEDURE

The following reagents are added to cuvettes with a critical volume of less than 1 ml. If a large number of assays is to be carried out, multiples of the indicated volumes of tris buffer and water may be mixed and appropriate quantities of the mixture distributed into cuvettes.

	Blank (μl)	System (μl)
Tris-HCl, 1 M, EDTA, 5 mM, pH 8.0	100	100
1:20 hemolyzate	20	20
H$_2$O	880	860
Incubate at 37° for 10 min		
Adenosine, 4 mM	—	20

The decrease of optical density of the system is measured against that of the blank at 265 nm at 37° C. A recorder expansion giving a full scale reading of 0.2 OD units is suitable for normal blood samples.

C. ADDITIONAL BLANKS REQUIRED

None.

D. COMMENTS, CALCULATION AND NORMAL VALUES

Because of the low activity of the enzyme and the necessity for measuring the OD at 265 nm, sensitive instrumentation is required. The difference in the mMolar extinction coefficient of adenosine and inosine is 8.10 (1). The equation on page 34 ($\epsilon = 8.10$) is used to calculate the results. Red cells of normal adults contain 1.11 ± 0.23 IU of adenosine deaminase/g Hb (mean $\pm$ S.D.). Factors which permit conversion of results obtained at 25° and 30° into values which would have been obtained with the same hemolyzate at 37° are presented in Appendix 2.

Adenosine deaminase deficiency has been associated, in several families, with immune deficiency disease.

REFERENCE

1. Bergmeyer, H. U.: Methods of Enzymatic Analysis. Academic Press, New York 1965, p. 493.

28

acetylcholinesterase

A. PRINCIPLE

Acetylcholinesterase catalyzes the hydrolysis of acetylthiocholine to thiocholine. The rate of production of thiocholine is measured by following the reaction of thiocholine with 5,5′-dithiobis(2-nitrobenzoic acid) (DTNB) which produces a yellow color due to the formation of 5-thio-2-nitrobenzoic acid. The rate of formation of the yellow anion is measured at 412 nm.

$$\text{Acetylthiocholine iodide} \xrightarrow{\text{acetylcholinesterase}} \text{Thiocholine} + \text{Acetate}$$

$$\text{Thiocholine} + \text{DTNB} \rightarrow \text{5-Thio-2-nitrobenzoic acid} + \text{Oxidized thiocholine}$$

B. PROCEDURE (based on Ellman et al. [1])

The following reagents are added to cuvettes with a critical volume of less than 1 ml. If a large number of assays is to be carried out, multiples of the indicated volumes of phosphate buffer, DTNB, and water may be mixed and appropriate quantities of the mixture distributed into cuvettes.

	Blank (μl)	System (μl)
Tris-HCl, 1 M, EDTA, 5 mM, pH 8.0	100	100
DTNB, 0.5 mM in 1% sodium citrate	50	50
1:20 hemolyzate diluted 1:10 in water	10	10
H$_2$O	840	790
Incubate at 37° C for 10 min		
Acetylthiocholine iodide, 10 mM	—	50

The increase in the optical density of the system is measured against that of the blank at 412 nm for 10 to 15 min.

A recorder expansion giving a full-scale reading of 1.0 OD unit is suitable for normal blood samples.

C. ADDITIONAL BLANKS REQUIRED

A blank is carried out in which β-mercaptoethanol-EDTA stabilizing solution is substituted for hemolyzate in both the blank and system mixtures and the optical density of the hemolyzate-free system is read against that of the hemolyzate-free blank at 412 nm.

D. COMMENTS, CALCULATIONS, AND NORMAL VALUES

The assay is carried out on whole hemolyzate for the sake of convenience, but acetylcholinesterase in red cells is entirely stroma-bound.

Although maximum activity is reached with the use of 1 mM acetylthiocholine iodide, 0.5 mM is used because the blank rate is lower at this concentration.

The blank rate should be subtracted from the rate measured in the presence of hemolyzate. The general equation on page 34 is used for calculating the activity of the enzyme, with $\epsilon = 13.6$.

The red cells of normal adults contain 36.9 ± 3.83 IU of acetylcholinesterase/g Hb (mean $\pm$ standard deviation). Factors which permit conversion of results obtained at $25°$ and at $30°$ into values which would have been obtained with the same hemolyzate at $37°$ are presented in Appendix 2.

Cholinesterase activity is markedly influenced by red cell age. Low red cell cholinesterase activity seems to be characteristic of paroxysmal nocturnal hemoglobinuria.

REFERENCE

1. Ellman, G. L., Courtney, K. D., Andres, V., Jr., and Featherstone, R. M.: A new and rapid colorimetric determination of acetylcholinesterase activity. Biochem. Pharmacol. 7:88–95, 1961.

29

catalase

A. PRINCIPLE

Catalase catalyzes the breakdown of H_2O_2 according to the following reaction:

$$2H_2O_2 \xrightarrow{\text{catalase}} 2H_2O + O_2$$

The rate of decomposition of H_2O_2 by catalase is measured spectrophotometrically at 230 nm, since H_2O_2 absorbs light at this wavelength.

Ethanol is added to stabilize the hemolyzate by breaking down "complex II" of catalase and H_2O_2.

B. PROCEDURE

The following reagents are added to cuvettes with a critical volume of less than 1 ml. If a large number of assays is to be carried out, multiples of the indicated volumes of tris buffer, H_2O_2, and water may be mixed and appropriate quantities of the mixture distributed into cuvettes.

	Blank (μl)	System (μl)
Tris-HCl, 1 M, EDTA, 5 mM, pH 8.0	50	50
H_2O_2, 10 mM*	—	900
H_2O	930	30

Incubate at 37° C for 10 min

| 1:2000 hemolyzate with ethanol† | 20 | 20 |

* Measure OD of 0.9 ml of a 1:10 dilution of 1 M phosphate buffer, pH 7.0, at 230 nm (OD_1). Add 0.1 ml of a 1:100 dilution of 30% H_2O_2 solution, and read OD (OD_2). Since the mM extinction coefficient of H_2O_2 at 230 nm is 0.071 [1], the H_2O_2 concentration (C) of the 1:100 diluted peroxide solution is 141 ($OD_2 - OD_1$) mM. To dilute to 10 mM for the assay, dilute 1 ml of the 1:100 dilution to $C/10$ ml with water.

† A 1:20 hemolyzate prepared in β-mercaptoethanol-EDTA stabilizing solution is diluted further 1:100, and 20 μl of absolute or 95% ethanol are added per milliliter of dilute hemolyzate to break down any "complex II" which may be present.

The decrease in optical density of the system is measured against that of the blank for 10 min.

A recorder expansion giving a full-scale reading of 1.0 OD unit is suitable for normal blood samples.

C. ADDITIONAL BLANKS REQUIRED

None.

D. COMMENTS, CALCULATIONS, AND NORMAL VALUES

The reaction is linear for only the first 3 or 4 min. Therefore, recording of optical density should begin immediately after the addition of hemolyzates.

The general equation on page 34 is used for calculating the activity of the enzyme, with $\epsilon = 0.071$.

The catalase reaction has first-order kinetics, so that expression of enzyme velocity as a first-order reaction constant is more appropriate than the use of enzyme units as calculated above. However, at the concentration of hydrogen peroxide and hemolyzate recommended, the amount of peroxide consumed is negligible compared with the total amount of peroxide available. Therefore, the units calculated are suitable, provided that the concentration of hydrogen peroxide in the reaction mixture is carefully controlled.

The red cells of normal adults contain $15.31 \times 10^4 \pm 2.39 \times 10^4$ IU of catalase/g Hb (mean $\pm$ standard deviation) at 37°. Factors which permit conversion of results obtained at 25° and at 30° into values which would have been obtained with the same hemolyzate at 37° are presented in Appendix 2.

Absence of red cell catalase occurs as a rare genetic defect. In addition, the disappearance of red cell catalase in the presence of aminotriazole can be used as a measure of the rate of hydrogen peroxide generation.

REFERENCE

1. Chance, B. Catalases and peroxidases, part II. Special Methods: Methods of Biochemical Analysis. 1:408–424, 1954.

30

galactose-1-phosphate uridyl transferase

Method 1. UDPG Consumption Method

A. PRINCIPLE

Galactose-1-phosphate uridyl transferase catalyzes the exchange reaction:

$$\text{UDPG} + \text{Galactose-1-P} \underset{\xleftarrow{\hspace{1cm}}}{\overset{\text{gal-1-PUT}}{\longrightarrow}} \text{UDPGal} + \text{Glucose-1-P}$$

In the UDPG consumption assay, the disappearance of UDPG from a mixture containing UDPG, galactose-1-P, and hemolyzate is estimated by measuring residual UDPG in the NAD-linked reaction catalyzed by UDPG dehydrogenase:

$$\text{UDPG} + 2\text{NAD}^+ \xrightarrow{\text{UDPG Deh.}} \text{UDP glucuronic acid} + \text{H}^+ + 2\text{NADH}$$

Because of the fact that the enzyme is not fully saturated with UDPG at the concentration at which it can be used in the consumption assay, and because of the inhibitory effect of the reaction product, UDP galactose (UDPGal), the reaction is not linear with hemolyzate concentration over a portion of the practical working range. An appropriate correction for nonlinearity can be made empirically (see Section D below).

Epimerase catalyzes an alternate route by which UDPG may be consumed in hemolyzates: its conversion to UDPGal. However, the activity of epimerase requires minute amounts of NAD to be present. Preincubation of hemolyzates results in destruction of endogenous NAD by NADase, inactivating the epimerase. In the case of newborn infants, very little red cell NADase may be present and preincubation with exogenous NADase may be required to inactivate epimerase.

91

B. PROCEDURE (based on Beutler and Baluda [1])

Washed red cells are mixed with an equal volume of cold distilled water, frozen and thawed, and incubated for 10 min at 37° C. If the blood is from an infant under 3 months of age, 0.025 parts of a solution of NADase (0.63 U/ml) are added to each part of hemolyzate prior to incubation.

The following reagents are mixed in the volumes indicated or in multiples. Galactose-1-P and UDPG-glycine may be premixed if a large number of assays is to be carried out.

	Blank (μl)	System (μl)
UDPG-glycine*	200	200
Galactose-1-P, 8 mM		100
H$_2$O	100	
Preincubate at 37° C for 3 to 5 min		
1:1 hemolyzate	200	200
Incubate at 37° C for exactly 15 min		
Ice-cold NaCl, 0.154 M (0.9%)	1000	1000

* Prepared daily by mixing one part of 7 mM UDPG with four parts of glycine buffer, 1 M, pH 8.7.

Place in boiling water bath, stirring intermittently, for 2 min. Centrifuge and remove supernatant.

Residual UDPG in the supernatant of the boiled extract is determined as follows.

	System (μl)
Glycine buffer, pH 8.7, 1 M	200
NAD, 10 mM	100
H$_2$O	400
Supernatant	200
UDPG dehydrogenase (800 Sigma U/ml = 0.032 IU/ml)	100

A recording spectrophotometer must be used, and the time at which the reaction is started with UDPG dehydrogenase is indicated on the chart, preferably using an auxiliary event marker. The full-scale expansion should be 1.0 OD unit. A rapid chart speed (1 inch/min) is used during the initial portion of the reaction, but then a slower speed ($\frac{1}{4}$ inch/min or slower) is quite suitable. The reaction is followed at 340 nm until there is no further change in optical density. There is no blank.

C. ADDITIONAL BLANKS REQUIRED

None.

D. COMMENTS, CALCULATIONS, AND NORMAL VALUES

The boiled extract used for UDPG measurements can be stored overnight at $-20°$ with very little loss of UDPG. Frozen UDPG dehydrogenase solution maintains activity for several weeks, but NADase is quite unstable and must be freshly prepared.

A large number of variations of the UDPG consumption assay have been described, but our results, and those of others [2], indicate that the version given above gives the most reproducible and linear data of any of the techniques. If sufficiently high concentrations of UDPG are used to give entirely linear results, a large experimentally derived value must be subtracted from a slightly larger one. This magnifies experimental errors. Low concentrations, which have been used in some assay procedures, result in marked deviation from linearity. Although in some published techniques the addition of sulfhydryl reagents has been advocated [3], this does not affect the activity of the red cell enzyme, only the activity of the white cell enzyme.

The concentration of UDPG in the filtrates is estimated by extrapolating the first portion of the progress curve backwards to the time at which the reaction was initiated. This optical density value is subtracted from the optical density at the completion of the reaction, when all of the UDPG has been oxidized. In the case of filtrates from the blank system, a stable value is usually reached after approximately 30 to 45 min. In some instances, a subsequent gradual decline is observed and the maximum value attained is used. In cuvettes containing filtrates from the complete system, a stable value is usually reached, also after about 30 to 45 min, but in some instances a small continuous rise in optical density is observed. If this occurs, the value after which no further change in rate is observed is used. This has been illustrated in detail elsewhere [1].

The activity of transferase (E) as micromoles of UDPG consumed per gram per hour is obtained as follows:

$$E = \frac{(\Delta OD_B - \Delta OD_R) \times 1210 \times F_c}{Hb}$$

where ΔOD_B is the change in optical density of the blank system, ΔOD_R is the optical density change of the reaction system, and Hb is the hemoglobin concentration of the hemolyzate in gram percent. This equation takes into account the various dilutions employed and the fact that 2 μmoles of NAD are reduced for each μmole of UDPG oxidized. If $(\Delta OD_B - \Delta OD_R)$ is greater than 0.200, an empirically derived correction factor (F_c) must be used to maintain linearity. The values of these factors are given in Table VI. To convert these units (micromoles per gram of Hb per hour) into international units per gram of Hb, it is necessary only to divide by 60, giving micromoles of UDPG

Table VI. Correction Factors for Galactose-1-P Uridyl Transferase Assays [1]

$\Delta OD_B - \Delta OD_R$	F_c	$\Delta OD_B - \Delta OD_R$	F_c
<0.200	1.000	0.265	1.132
0.200	1.010	0.270	1.144
0.205	1.020	0.275	1.160
0.210	1.029	0.280	1.175
0.215	1.037	0.285	1.189
0.220	1.045	0.290	1.203
0.225	1.053	0.295	1.220
0.230	1.061	0.300	1.243
0.235	1.072	0.305	1.269
0.240	1.083	0.310	1.294
0.245	1.094	0.315	1.317
0.250	1.100	0.320	1.353
0.255	1.110	0.325	1.412
0.260	1.119	0.330	1.479

consumed per minute per gram of hemoglobin. However, it has been traditional to use micromoles of UDPG consumed per hour as a basis of measurement of this enzyme.

The measurement of galactose-1-P uridyl transferase activity is of interest primarily for the detection of galactosemia and its carrier state.

It is difficult to present normal values for this enzyme, because several subactive genetic variants are prevalent in the population. The normal range for the enzyme appears to be approximately 18.5 to 28.5 IU/g Hb galactose-1-P uridyl transferase. Subjects heterozygous for the Duarte variant have activity in the range or 13.5 to 18.5 units, while heterozygotes for galactosemia and homozygotes for the Duarte variant have activity in the 8.5 to 13.5 range. Individuals doubly anomalous for the gene for galactosemia and the gene for the Duarte variant will have activity between 3.5 and 8.5 units. With rare exceptions, no detectable activity is present in the red cells of individuals with galactosemia.

Method 2. Fluorometric Method

A. PRINCIPLE

In this technique, the rate of the transferase reaction is followed by measuring the formation of glucose-1-P:

$$\text{UDPG} + \text{galactose-1-P} \xrightarrow{\text{gal-1-PUT}} \text{UDPGal} + \text{glucose-1-P}$$

The endogenous phosphoglucomutase of the hemolyzate, together with en-dogenous glucose-1,6-diP, results in the conversion of glucose-1-P to glu-cose-6-P. The NADP in the reaction mixture is reduced to NADPH in the course of the glucose-6-P dehydrogenase and 6-phosphogluconate dehydro-genase reactions. The enzymes catalyzing these two steps are present in excess in the vast majority of hemolyzates. In the case of severe glucose-6-P dehydro-genase deficiency, as occurs in Mediterranean subjects (but not in the milder, Negro A- type deficiency), it may be necessary to add exogenous glu-cose-6-P dehydrogenase.

After the reaction mixture has been incubated with the blood sample, the reaction is stopped by dilution and the NADPH formed is measured fluoro-metrically. Using the same diluted sample, the hemoglobin concentration is estimated, using the intensely absorbing Soret band at 410 nm.

B. PROCEDURE (based on Beutler and Mitchell [4])

The reaction mixture is prepared in bulk using the following volumes or multiples thereof. It may be stored frozen for about 1 month.

	Volume (ml)
UDPG, 10 mM	0.6
Galactose-1-P, 27 mM	0.6
NADP, 6 mM	0.8
Tris-acetate, pH 8.0, 0.75 M	2.0
Saponin (Sigma), 1%	0.8
Disodium EDTA, 27 mM	0.09
MgCl$_2$, 0.1 M	0.13
H$_2$O	1.00
(Glucose-6-P dehydrogenase, 30 U/ml	0.05)

The addition of glucose-6-phosphate dehydrogenase is optional. It is usu-ally not required, but may be included if severe G-6-PD deficiency may be present in the samples to be tested.

Ten microliters of whole blood are added to 100 μl of reaction mixture. After 30 min incubation at 37° C, 20 μl of the mixture are diluted in 4 ml of potassium phosphate buffer, 0.01 M, pH 7.4. Fluorescence measurements are made in a 10-mm (inner diameter) cuvette against a blank prepared by adding 20 μl of reaction mixture without blood to 4 ml of the phosphate buffer, using a Corning 7-60 primary and 3-72 secondary filter. The diluted reaction mixture is then transferred to a suitable photoelectric colorimeter cuvette, and the optical density is measured at 410 nm against a buffer blank to estimate the hemoglobin content.

C. ADDITIONAL BLANKS REQUIRED

None.

D. COMMENTS, CALCULATIONS, AND NORMAL VALUES

The amount of fluorescence produced depends on a number of variables, including the electronic characteristics of the detecting instrument and the intensity of the light source. The latter, in particular, is subject to change with time. Accordingly, calibration must be carried out periodically.

Galactose-1-P uridyl transferase activity (E) in micromoles of UDPG consumed per gram of Hb per hour is obtained as follows:

$$E = \frac{FC}{A}$$

where F is the fluorescence reading of the sample, C is the calibration factor, and A is the absorbance of the diluted system at 410 nm.

The calibration factor (C) must be determined for each fluorometer and photoelectric colorimeter used in the procedure. A solution of NADPH containing approximately 1 mg/ml is prepared, and its actual concentration is measured as outlined on pages 20–21. A series of dilutions in 0.01 M potassium phosphate buffer, pH 7.4, containing 0.25 to 1.25 μM NADPH is prepared, and fluorescence readings are made against the buffer blank. The readings are plotted against the NADPH concentration on arithmetic graph paper. The fluorescence reading given by 1 μM NADPH is read from the curve and designated F_1.

Calibration of the photoelectric colorimeter is carried out by hemolyzing a normal blood sample by freezing and thawing and then determining its hemoglobin concentration with ferricyanide-cyanide reagent (page 11). One-tenth milliliter of blood is then added to 1 ml of a 0.133% saponin solution. Further dilutions of 1:100, 1:200, and 1:400 are prepared in 0.01 M potassium phosphate buffer, pH 7.4, and the optical density of each dilution is determined at 410 nm against a buffer blank. The concentration of hemoglobin in grams per liter is plotted against the optical density of each dilution. A slight deviation from Beer's law can be observed with some colorimeters, but this will not be sufficiently great to introduce a serious error, and the best fit is employed. The optical density given by a 0.1 g/liter solution of hemoglobin is read from the line and is designated A_1. The calibration factor (C) may now be calculated:

$$C = \frac{10 \times A_1}{F_1}$$

The results obtained in this assay procedure are similar to those obtained with the UDPG consumption assay. The red cells of normal adults contain 28.40 ± 6.94 IU of galactose-1-P uridyl transferase/g Hb (mean $\pm$ standard deviation). There is a tendency for the value to be somewhat higher than those observed in the consumption assay because the concentration of UDPG used is greater. Thus the enzyme is more completely saturated with the substrate, and the competitive inhibition exerted by the reaction product, UDPGal, is negligible.

REFERENCES

1. Beutler, E., and Baluda, M. C.: Improved method for measuring galactose-1-phosphate uridyl transferase activity of erythrocytes. Clin. Chim. Acta 13:369–379, 1966.
2. Ellis, G., and Goldberg, D. M.: The enzymological diagnosis of galactosaemia. Ann. Clin. Biochem. 6:70–73, 1969.
3. Mellman, W. J., and Tedesco, T. A.: An improved assay of erythrocyte and leukocyte galactose-1-phosphate uridyl transferase. Stabilization of the enzyme by a thiol protective reagent. J. Lab. Clin. Med. 66:980–986, 1965.
4. Beutler, E., and Mitchell, M.: New rapid method for the estimation of red cell galactose-1-phosphate uridyl transferase activity. J. Lab. Clin. Med. 72:527–532, 1968.

31

galactokinase (GK)

A. PRINCIPLE

Galactokinase catalyzes the reaction

$$\text{Galactose} + \text{ATP} \xrightarrow[\text{Mg}^{2+}]{\text{GK}} \text{Galactose-1-P} + \text{ADP}$$

In this assay $1\text{-}^{14}\text{C-galactose}$, which is incubated with ATP and hemolyzate, is used as substrate. The reaction is stopped by dilution with a large excess of nonradioactive galactose. The 1-^{14}C-galactose-1-P formed is separated from the nonphosphorylated radioactive galactose by spotting on DEAE paper, which binds the phosphorylated galactose, followed by washing with distilled water to remove the nonphosphorylated galactose.

B. PROCEDURE (based on Sherman and Adler [1] and Beutler and Matsumoto [2])

It is necessary to purify commercially available 1-^{14}C-galactose to remove traces of sugars which could be phosphorylated by hexokinase, and to remove labeled contaminants which adhere to DEAE and would, therefore, give high blank values. A small column is prepared by placing a little glass wool into the bottom of a tuberculin syringe barrel, and adding microgranular DEAE (Whatman DE-52) slurry to give a volume of 0.2 to 0.3 ml. The column is washed with a few drops of 3.8 mM galactose solution followed by 5 ml of water. 1-^{14}C-Galactose is diluted to contain approximately 10 μCi/ml. Two microliters of 1 M tris-HCl buffer, pH 8, 5 μl of 0.1 M MgCl$_2$, 100 μl of hexokinase diluted in water to contain 2 U/ml, and 10 μl of 60 mM neutralized ATP are added to each milliliter of the ^{14}C-galactose solution. After 60 min incubation at 37° C, the reaction mixture is passed through the column, which is then washed with sufficient additional distilled water to give five times the original volume. The radioactive galactose solution which has been collected contains approximately 2 μCi of ^{14}C per milliliter and is stable when frozen.

A partial reaction mixture is prepared by mixing the following reagents in the proportion shown:

	μl
NaF, 100 mM	250
MgCl$_2$, 100 mM	500
Tris-HCl, 1 M, pH 7.4	1000
Galactose, 7.6 mM	250
^{14}C-Galactose, 2 μCi/ml	500
ATP, 0.12 M	250
Saponin (Sigma), 1%	1000
H$_2$O	250

Blood can be collected either in EDTA or ACD solution. Fifty microliters of the whole blood are added to 200 μl of the reaction mixture. Two 50-μl aliquots of the reaction mixture are removed immediately and mixed on a spot plate or in a small test tube with 20 μl of a 1 M galactose solution. Fifty microliters of each mixture are then spotted on a circle of DEAE paper (Whatman DE 81) 15 to 20 mm in diameter. One of the spotted papers is dropped immediately into a small cup of water (0 time sample), while the other is permitted to dry (100% standard). The remainder of the reaction mixture is permitted to incubate for 60 min at 37°, and another sample is treated in a fashion identical with the 0 time sample. This is designated as the 60-min sample. The 0 time sample and the 60-min sample are placed, while wet, on a sintered glass funnel and are washed with 600 to 800 ml of distilled water which is drawn through the funnel with a vacuum. These two samples are now permitted to dry and are ready for counting. When multiple samples are assayed, the estimation of the radioactivity of the 0 time sample and of the 100% standard need only be carried out on a single blood sample, since these will vary little from one sample to the next.

The dried samples may be pasted onto planchettes and counted with a gas-flow counter, or, more efficiently, may be placed in a scintillation solution, such as toluene with 0.3 g POPOP and 5 g PPO per liter, and the radioactivity of each sample is determined.

C. ADDITIONAL BLANKS REQUIRED

No additional blanks are needed. The 0 time sample is used rather than a blank without ATP, since the endogenous ATP content of the hemolyzate will result in some galactose phosphorylation.

D. COMMENTS, CALCULATIONS, AND NORMAL VALUES

It is particularly important that freshly prepared galactose solutions not be used in the assay. When pure α- or β-D-galactose is dissolved in water, several reactions take place. α-Galactose and β-galactose are interconverted until an equilibrium mixture containing approximately 30% α-galactose and 70% β-galactose is formed. In addition to these pyranose forms, furanose forms of galactose are also produced. Finally, there is formation of a keto sugar, presumably tagatose. The last transformation is quite slow. Freshly prepared galactose solutions should therefore be incubated at 37° for at least 4 hours prior to use, to facilitate equilibrium between reacting and nonreacting forms of galactose, or aged for 3 or 4 days at 4° C before they are used.

The impurity of commercially obtained ^{14}C-galactose solutions represents a major source of difficulty in this assay. The purification procedure given above has proved to be very satisfactory. However, we have also encountered preparations from more than one manufacturer which were labeled as ^{14}C-galactose but which contained little or no galactose. Substantial dilution of ^{14}C-galactose with impurities will produce a major inaccuracy in the assay. When a new preparation of ^{14}C-galactose is purchased, it is useful to assay the substrate, after preliminary purification, for galactose content. This may be done by preparing a partial reaction mixture, as used for galactokinase assay, deleting, however, the unlabeled galactose. If 100 μl of such a reaction mixture are incubated with 100 μl of 1:2 hemolyzate, preferably from cord blood, for several hours, 70 to 85% of the galactose should be phosphorylated. We have not observed complete phosphorylation, probably because some of the ^{14}C-galactose is in the form of slowly interconverting, nonreacting forms of galactose.

The activity of galactokinase (E) in micromoles of galactose phosphorylated per minute per gram of Hb is obtained as follows:

$$ E = \frac{(A_T - A_0) \times C}{A_{100\%} - \text{bg}} \times \frac{1}{T \times 0.01 \text{ Hb} \times 0.2} $$

where A_T is the radioactivity of the sample after T minutes incubation, A_0 is the radioactivity of the 0 time sample, $A_{100\%}$ is the 100% standard, bg is the background, C is the concentration of galactose (in micromoles per milliliter) in the final assay system, and Hb is the concentration of hemoglobin in grams percent. The factor 0.01 is required to convert the latter figure to grams per milliliter in the initial hemolyzate, and 0.2 is needed to convert this to the Hb concentration in the final system. In the assay described, $T = 60$ min.

$$ C = C_1 + \frac{R}{S} $$

where C_1 is the concentration of nonradioactive galactose in the final reaction mixture in micromoles per milliliter (0.38 in the system described here), S is the specific activity of the radioactive galactose in microcuries per micromole, and R is the radioactivity in the assay system in microcuries per milliliter (0.2 in the system as described). It is usually more convenient to describe activity of the enzyme in terms of 1000E, or milliunits of activity per gram of Hb. Under the conditions of the assay, as presented above, the activity of the enzyme, in milliunits per gram of Hb, is

$$ \frac{8333(A_T - A_0) \times (0.38 + 0.2/S)}{(A_{100} - \text{bg}) \times \text{Hb}} $$

The red cells of normal adults contain 29.12 $\pm$ 5.97 mU of galactokinase activity per gram of Hb (mean $\pm$ standard deviation) at 37°. The red cells of infants have three to four times the adult level of galactokinase activity.

Homozygous galactokinase deficiency causes a form of galactosemia which is associated with cataracts, but not with mental deficiency or liver disease. Heterozygotes, who have approximately half of the normal enzyme activity, may also have increased susceptibility to cataracts in young adult life.

REFERENCES

1. Sherman, J. R., and Adler, J.: Galactokinase from *Escherichia coli.* J. Biol. Chem. 238:873–878, 1963.
2. Beutler, E., and Matsumoto, F. T.: A rapid simplified assay for galactokinase activity in whole blood. J. Lab. Clin. Med. 82:818–821, 1973.

32

UDP glucose-4-epimerase (epimerase)

A. PRINCIPLE

The formation of UDPG can be followed spectrophotometrically by coupling it to the enzyme UDPG dehydrogenase in the reaction.

$$\text{UDPGal} \underset{\text{NAD}}{\overset{\text{epimerase}}{\rightleftharpoons}} \text{UDPG}$$

The formation of UDPG can be followed spectrophotometrically by coupling it to the enzyme UDPG dehydrogenase in the reaction

$$\text{UDPG} + 2\text{NAD}^+ \rightarrow \text{UDP glucuronate} + 2\text{NADH} + 2\text{H}^+$$

The reduction of NAD is followed at 340 nm.

B. PROCEDURE

The following reagents are added to cuvettes with a critical volume of less than 1 ml. If a large number of assays is to be carried out, multiples of the indicated volumes of tris buffer, NAD, and water may be mixed and appropriate quantities of the mixture distributed into cuvettes.

	Blank (μl)	System (μl)
Tris-HCl, 1 M, EDTA, 5 mM, pH 8.0	100	100
NAD, 10 mM	100	100
UDPG dehydrogenase* (800 Sigma		
U/ml = 0.032 IU/ml)	500	500
H$_2$O	250	200
1:20 hemolyzate	50	50
Incubate at 37° C for 10 min		
UDPGal, 5 mM	0	50

* Dissolve and centrifuge for 10 min; use clear supernatant in assay.

The increase in the optical density of the system is measured against that of the blank at 340 nμm at 37° C. A recorder expansion giving a full-scale reading of 0.1 OD unit is suitable for normal blood samples.

C. ADDITIONAL BLANKS REQUIRED

A blank assay should be carried out to be certain that the UDPG dehydrogenase used is free of epimerase activity. β-Mercaptoethanol-EDTA stabilizing solution is substituted for hemolyzate in both the blank and system mixtures, and the change in optical density of the hemolyzate-free system is measured against that of the hemolyzate-free blank.

If no change in optical density is observed, indicating that the UDPG dehydrogenase is free of epimerase activity, it is not necessary to carry out the blank determination each time, as long as the same UDPG dehydrogenase preparation is used. If some residual activity is present in the auxiliary enzyme, however, this blank should be determined each day.

D. COMMENTS, CALCULATIONS, AND NORMAL VALUES

This is the least active of all of the enzymes which we assay spectrophotometrically. Thus very sensitive instrumentation is required and technical variability is unusually great. It is possible that a fluorometric modification of this assay would be more satisfactory, but details of such a procedure have not yet been worked out.

The blank rate, if any (see Section C above), should be subtracted from the rate measured in the presence of hemolyzate. Two moles of NAD are reduced for each mole of UDPGal converted to UDPG. The equation on page 34 (ϵ = 12.44) is used.

The red cells of normal adults contain 0.231 ± 0.061 IU of epimerase/g Hb (mean ± standard deviation) at 37°. Factors which permit conversion of results obtained at 25° and at 30° into values which would have been obtained with the same hemolyzate at 37° are presented in Appendix 2.

One patient with epimerase deficiency has been described in the literature.

part V

intermediate compounds

33

ATP

A. PRINCIPLE

ATP phosphorylates glucose in the hexokinase (Hx) reaction:

$$\text{Glucose} + \text{ATP} \xrightarrow[\text{Mg}^{2+}]{\text{Hx}} \text{Glucose-6-P} + \text{ADP}$$

The glucose-6-phosphate formed is measured in the glucose-6-phosphate dehydrogenase reaction:

$$\text{Glucose-6-P} + \text{NADP}^+ \xrightarrow{\text{G-6-PD}} \text{6-PGA} + \text{NADPH} + \text{H}^+$$

The change in optical density at 340 nm, resulting from the reduction of NADP to NADPH, is measured, and, in the presence of excess glucose, represents the amount of ATP in the mixture.

B. PROCEDURE

The estimation may be carried out on extracts prepared from whole blood, since almost all of the blood ATP is in the red cells. A hemoglobin estimation is carried out on the whole blood sample by adding 20 μl of blood to 10 ml of ferricyanide-cyanide reagent, and measuring the optical density at 540 nm. To prepare the extract for ATP determination, 1 ml of ice-cold 20% perchloric acid is added to 2 ml of ice-cold whole blood and mixed well. One milliliter of the supernatant solution after centrifugation is pipetted into a graduated tube, neutralized with 3 M K_2CO_3 solution using methyl orange indicator (see page 16), and the volume is adjusted to 1.5 ml. The following reagents are added to cuvettes with a critical volume of less than 1 ml. If a large number of assays is to be carried out, multiples of the indicated volumes of tris buffer, glucose, NADP, $MgCl_2$, G-6-PD, and water may be mixed and appropriate quantities of the mixture distributed into cuvettes.

	Blank (μl)	System (μl)
Tris-HCl, 1.0 M, EDTA, 5 mM, pH 8.0	100	100
MgCl$_2$, 0.1 M	20	20
NADP, 2 mM	200	200
Glucose, 20 mM	50	50
PCA extract	—	200
H$_2$O	615	415
G-6-PD, 60 U/ml (diluted in β-mercaptoethanol-EDTA stabilizing solution)	5	5

Read baseline at 340 nm at 37° C

| Hexokinase, 400 U/ml | 10 | 10 |

Read OD until constant value is reached

A recorder expansion giving a full-scale reading of 1 OD unit is suitable for normal blood samples.

C. ADDITIONAL BLANKS REQUIRED

None.

D. COMMENTS, CALCULATIONS, AND NORMAL VALUES

The same extract used for ATP determinations can also be used for the determination of AMP and ADP. If the extract is to be stored, it should be neutralized prior to freezing.

The ATP concentration, in micromoles of ATP per gram of Hb (C), is calculated as

$$C = \frac{\Delta OD \times 0.313}{OD_{540} \times F_{HB}}$$

where ΔOD is the change in optical density at 340 nm which occurs after addition of hexokinase; OD_{540} is the optical density at 540 nm when 0.02 ml of the blood has been added to 10 ml of ferricyanide-cyanide solution; and 0.313 is a combined factor obtained from the extinction coefficient of NADPH, the dilution of extract in the cuvette, the conversion of hemoglobin from grams percent to grams per milliliter, and from the dilution of the original sample, where it is assumed that the water content of blood is 80% (see page 13).

The level of red cell ATP is of particular interest in the study of blood preservation, and has also been investigated in many red cell disorders. It is correlated with the inorganic phosphate level of serum. ATP levels are somewhat higher in the red cells of Caucasians than in those of Negroes. The red cells of normal Caucasian adults contain 4230 ± 290 nmoles of ATP/g Hb (mean $\pm$ standard deviation); those of Negro adults contain 3530 ± 301 nmoles of ATP/g Hb.

34

AMP and ADP

A. PRINCIPLE

ADP is measured in a perchloric acid extract of red cells through the reactions

$$PEP + ADP \xrightarrow[\text{Mg}^{2+}]{\text{PK}} \text{Pyruvate} + ATP$$
$$\text{Pyruvate} + NADH + \underset{}{\overset{\text{LDH}}{\rightleftharpoons}} \text{Lactate} + NAD^+$$

The amount of pyruvate formed bears a stoichiometric relationship to the amount of ADP available, and hence the oxidation of NADH to NAD measured at 340 nm corresponds to the amount of ADP in the filtrate.

The addition of adenylate kinase converts AMP into ADP in the reaction

$$AMP + ATP \underset{\text{Mg}^{2+}}{\overset{\text{AK}}{\rightleftharpoons}} 2ADP$$

The ADP formed then serves as substrate in the pyruvate kinase reaction and can be measured through a further decrease of optical density at 340 nm as NADH is oxidized to NAD.

B. PROCEDURE

The same extract used for assay of ATP (see Chapter 33) may be used. The assay is carried out as follows.

	Blank (μl)	System (μl)
Tris-HCl, 1 M, EDTA, 5 mM, pH 8.0	50	50
MgCl$_2$, 0.1 M	20	20
PCA extract	—	700
H$_2$O	705	5
PEP, 0.015 M	50	50
NADH, 2 mM	100	100
Lactate dehydrogenase, 240 U/ml	50	50
ATP (neut.), 20 mM	5	5

Read OD at 340 nm

Pyruvate kinase (Sigma, type II), 140 U/ml	10	10

Read OD at 340 nm until a constant value is reached

Adenylate kinase (myokinase), 725 U/ml	10	10

Read OD at 340 nm until a constant value is reached

A recorder expansion giving a full-scale reading of 0.4 OD unit is suitable for normal blood samples. When using a recording spectrophotometer which does not have a provision for arbitrarily adjusting the position of the recorder pen, or when using a nonrecording spectrophotometer, the instrument should be blanked at an optical density value of 0.35, since the optical density of the system will fall.

C. ADDITIONAL BLANKS REQUIRED

None.

D. COMMENTS, CALCULATIONS, AND NORMAL VALUES

The optical density should be relatively stable before the addition of pyruvate kinase. An excessively rapid baseline rate may be observed if the lactic dehydrogenase preparation is contaminated with pyruvate kinase. Certain crude pyruvate kinase preparations (e.g., Sigma, type I) have been found to be unsatisfactory for this assay. Care must be taken to avoid overtitration of the PCA, since a very alkaline filtrate will inhibit the enzymatic reactions involved.

The decrease in optical density after addition of pyruvate kinase represents the oxidation of 1 mole of NADH for each mole of ADP in the system. The concentration of ADP, in micromoles per gram of Hb (C), is obtained as follows:

$$C = \frac{\Delta OD \times 0.0896}{OD_{540} \times F_{HB}}$$

where ΔOD is the change in optical density at 340 nm after the addition of pyruvate kinase, F_{HB} is the hemoglobin factor (see page 11), OD_{540} is the optical density at 540 nm of 10 ml of ferricyanide-cyanide reagent after addition of 0.020 ml of the blood sample, and 0.0896 is a combined factor obtained from the extinction coefficient of NADH, the dilution of filtrate in the cuvette, dilution of blood for hemoglobin determinations, the conversion of hemoglobin from grams percent to grams per milliliter, and from the dilution of the original sample, where it is assumed that the water content of blood is 80% (see page 13). The change of OD after the addition of adenylate kinase represents the AMP in the reaction system, 1 mole of AMP resulting in the oxidation of 2 moles of NADH. Thus the concentration of AMP in micromoles per gram of Hb (C) is obtained as follows:

$$C = \frac{\Delta OD \times 0.0447}{OD_{540} \times F_{HB}}$$

The red cells of normal adults contain 635 ± 105 nmoles of ADP/g Hb (mean $\pm$ standard deviation). The red cells of normal adults contain 62 ± 10 nmoles of AMP/g Hb (mean $\pm$ standard deviation).

35

2,3-diphosphoglycerate (2,3-DPG)

A. PRINCIPLE

The assay depends on the requirement of the monophosphoglycerate mutase (MPGM) reaction for 2,3-DPG. Phosphoenolpyruvate (PEP) serves as a source of 2-PGA through the enolase reaction.

$$PEP \underset{}{\overset{\text{enolase}}{\rightleftharpoons}} 2\text{-PGA} \underset{\text{2,3-DPG}}{\overset{\text{MPGM}}{\rightleftharpoons}} 3\text{-PGA}$$

The concentrations of PEP and 2-PGA are permitted to reach equilibrium. When 2,3-DPG is present, 2-PGA is converted into 3-PGA, displacing the PEP/2-PGA equilibrium. This results in the conversion of PEP to 2-PGA.

The disappearance of PEP from the reaction mixture is measured spectro-photometrically at 240 nm.

B. PROCEDURE (based on Krimsky [1] and Beutler et al. [2])

Two-tenths milliliter of whole blood is added to 2 ml of ice-cold distilled water. A 0.2-ml aliquot of the hemolyzate is transferred to ferricyanide-cyanide reagent for hemoglobin estimation. A 1:100 dilution of the hemolyz-ate is made in a 1:100 dilution of tris-HCl, pH 7.4, 2 M. This dilution is suitable for assay on the same day if kept at 4° C, or for several days if frozen. If it is to be saved longer, it may be placed in a boiling water bath for 10 min and then frozen. The small amount of coagulum which forms will settle to the bottom, and the supernatant can be used directly for assay. In the case of samples with very low activity, 0.2 ml of whole blood is added to 0.8 ml of ice-cold distilled water, and 1 ml of cold 2% NaCl solution is mixed with the hemolyzate. A hemoglobin estimation is carried out on a 0.2-ml aliquot of the hemolyzate, and the remainder is placed in a boiling water bath for 10 min. The supernatant after centrifugation is then used for assay. A stock standard solution containing 1 mM 2,3-DPG in water is pre-pared. It is stable for at least several months at $-20°$ C.* Prior to assay,

---

* The stock solution should be assayed periodically for 2,3-DPG content. This is best done using a modification of the method described by Keitt [3] in which 2,3-DPG is converted to 3-PGA through the action of the phosphatase activity of commercially available monophosphoglyceromutase. This method may also be used to assay the 2,3-DPG content of perchloric acid extract of red cells, but the method described in this chapter is more suitable for the performance of large numbers of assays on small volumes of blood. The following method is used for assay of the standard:

	Blank (μl)	System (μl)
Tris-HCl, 1 M, EDTA, 5 mM, pH 8.0	100	100
KH$_2$PO$_4$, 1 M	5	5
MgCl$_2$, 0.1 M	100	100
GSH, 0.1 M	20	20
Hydrazine SO$_4$, 0.2 M	50	50
NADH, 2 mM	50	50
ATP, 20 mM	100	100
2,3-DPG standard 1 mM	—	50
PGK, 200 U/ml	5	5
GAPD, 680 U/ml	5	5
H$_2$O	515	465

Obtain baseline reading at 340 nm

10 mM Phosphoglycolate with		
MPGM, 550 U/ml	50	50

Read until no further change in OD 340

A 1 mM solution of 2,3-DPG will produce a 0.311 fall in the optical density of the system, when compared with the blank. The concentration of the standard, (C), then, is ΔOD/0.311 mM.

a 1 : 1000 dilution of the standard is made in a 1 : 100 dilution of 2 *M* tris-HCl, pH 7.4.

A reaction mixture is prepared daily by mixing the following volumes of reagents or multiples thereof and is stored at 0° C.

	Volume (ml)
Tris-HCl, 2 *M*, pH 7.4	0.4
$MgCl_2$, 0.5 *M*	0.2
PEP, 0.025 *M*	0.6
EDTA (neut.), 0.27 *M*	0.2
Monophosphoglycerate mutase, ~1600 U/ml	0.2
Enolase, ~32 U/ml	0.2
Bovine serum albumin, 5 mg/ml	1.8
H_2O	14.4

It is imperative to prepare a sufficient volume of this mixture for the entire day's work.

Nine-tenths milliliter of the reaction mixture is placed in each cuvette and allowed to stand at room temperature for 10 min. Sufficient water and diluted hemolyzate or extract or standard solution are added to give a total volume of 1.000 ml, and the change in optical density is recorded at 240 nm at 25° C against an air or water blank. Standards of 0, 25, 50, 75, and 100 μl are measured at the beginning and end of the day. Ten or twenty microliters of the 1 : 100 diluted hemolyzate are adequate when samples with the normal 2,3-DPG activity are examined. A recorder expansion giving a full-scale reading of 1.0 OD unit is suitable for normal blood samples.

C. ADDITIONAL BLANKS REQUIRED

None.

D. COMMENTS, CALCULATIONS, AND NORMAL VALUES

The rate of the reaction will depend not only on the 2,3-DPG concentration but also on the activity of the monophosphoglyceromutase in the reaction mixture. There is considerable variation between the activities and mode of assay of various preparations of the enzyme, and it may therefore be necessary to adjust the concentration of this material so that the optical density change obtained with the 50-μl standard is approximately 0.014 to 0.018 OD U/min. The assay procedures used by different manufacturers (and indeed the same manufacturer at different times) will also vary, so that the suggested concentration of phosphoglyceromutase should be regarded only as a very rough guide.

For reasons which are not entirely clear, the results are more reproducible at 25° C than at the usual assay temperature of 37° C.

The rate of change in optical density at 240 nm for each of the standards is plotted against the concentration of 2,3-DPG in the cuvette. This will be 25 mμM for the cuvette containing 25 μl of standard, 50 mμM for the cuvette containing 50 μl of standard, and so on. It is important to include in the calibration curve the standard without any added 2,3-DPG (0 mμM 2,3-DPG). The concentration of 2,3-DPG for each of the cuvettes to which extracts from samples to be tested have been added is then read from the curve. Calculation of 2,3-DPG concentration (C) in micromoles of 2,3-DPG per gram of Hb can then be made from the equation:

$$C = \frac{DR}{V_\text{H} \times \text{Hb} \times 10{,}000}$$

where D is the dilution made of the first hemolyzate (100 for a 1:100 dilution, etc.), V_H is the volume (in milliliters) of extract added to the cuvette, R is the concentration of 2,3-DPG in millimicromolar as read from the calibration curve, and Hb is the concentration in grams percent of hemoglobin in the hemolyzate prior to boiling. The factor 10,000 is obtained because it is necessary to multiply the hemoglobin concentration in grams percent by 10 to convert it to a value in grams per liter. Since the calibration curve gives a concentration of millimicromoles per liter, the values must also be divided by 1000 to give the concentration in terms of micromoles per liter.

Red cell 2,3-DPG is important in regulating the position of the oxygen dissociation curve of hemoglobin. Normal red cells contain 12270 ± 187 nmoles of 2,3-DPG/g Hb (mean ± standard deviation). Red cell 2,3-DPG levels increase in anemia and decrease rapidly in the storage of blood, particularly in acid media. Red cell 2,3-DPG levels are uniformly increased in patients with pyruvate kinase deficiency.

REFERENCES

1. Krimsky, I.: D-2,3-Diphosphoglycerate. In Bergmeyer, H. U. (Ed.), Methods of Enzymatic Analysis. New York, Academic Press, 1965, pp. 238–240.
2. Beutler, E., Meul, A., and Wood, L. A.: Depletion and regeneration of 2,3-diphosphoglyceric acid in stored red blood cells. Transfusion 9:109–114, 1969.
3. Keitt, A. S.: Reduced nicotinamide adenine dinucleotide-linked analysis of 2,3-diphosphoglyceric acid: Spectrophotometric and fluorometric procedures. J. Lab. Clin. Med. 77:470–475, 1971.

36

reduced glutat ,one (GSH)

A. PRINCIPLE

Virtually all of the nonprotein sulfhydryl of red cells is in the form of reduced glutathione (GSH). 5,5'-Dithiobis(2-nitrobenzoic acid) (DTNB) is a disulfide compound which is readily reduced by sulfhydryl compounds, forming a highly colored yellow anion. The optical density of this yellow substance is measured at 412 nm.

B. PROCEDURE (based on Beutler et al. [1])

Two-tenths milliliter of whole blood is added to 2.0 ml of distilled water, 0.2 ml of the lyzate is added to 10 ml of ferricyanide-cyanide reagent for hemoglobin estimation, and 3 ml of precipitating solution* are added to the remaining 2 ml of hemolyzate. After standing for 5 min, the mixture is filtered through a medium or coarse grade of filter paper. Two milliliters of filtrate are added to 8 ml of 0.3 M Na_2HPO_4 solution in a cuvette with a critical volume of 10 ml or more. It is read at 412 nm against a blank prepared by adding 2 ml of 2:5 water-diluted precipitating solution* (sham filtrate) to 8 ml of phosphate solution. A second optical density reading is taken after 1 ml of DTNB reagent** has been added to the blank cuvette and the cuvette containing filtrate.

C. ADDITIONAL BLANKS REQUIRED

No additional blanks are needed. If samples are to be tested repeatedly throughout the day, it is convenient to read the blank containing sham filtrate against a cuvette containing water. All subsequent samples are then read

* One hundred milliliters contain 1.67 g of glacial metaphosphoric acid, 0.2 g of disodium EDTA, and 30 g of sodium chloride.
** Twenty milligrams of DTNB per 100 ml of 1% sodium citrate solution.

against this water blank, and the reading of the sham filtrate blank is subtracted from all of the sample readings.

D. COMMENTS, CALCULATIONS, AND NORMAL VALUES

The entire procedure may be carried out at room temperature. However, the red cell lyzate should not be permitted to stand for more than a few minutes, since GSH will gradually be oxidized. The GSH in the acid filtrate is considerably more stable, even at room temperature, and readings may be delayed for several hours.

The precipitating solution will be cloudy because of undissolved EDTA, but this has no adverse effect on the results. The precipitating solution is stable for several weeks in the refrigerator, but will gradually lose its effectiveness. It may be used as long as a clear filtrate is obtained. Up to 0.5 ml of blood may ordinarily be used. Washed cells will give identical results. Full color development is observed immediately after the addition of the DTNB reagent. The color will fade very slowly, and readings within the first 10 min are quite satisfactory.

The molar extinction coefficient of the yellow anion produced when GSH interacts with DTNB is 13,600. However, when a band width greater than 6 nm is used, a lower extinction coefficient is obtained. A derived extinction coefficient E_1, which corrects both for differences in light path and band width, may be obtained by carrying out a GSH determination and reading an aliquot in a spectrophotometer with a narrow slit in a 1-cm cuvette to obtain the extinction value D_1. An aliquot of the same sample is then read in the system being calibrated, obtaining a second optical density reading D_2.

The correction factor, E_1, is obtained as follows:

$$E_1 = \frac{D_1}{D_2}$$

In the Coleman Jr. spectrophotometer we have consistently obtained E_1 values of 0.542 with a 24 mm cuvette. The concentration of glutathione in micromoles per gram of Hb (C) is obtained as follows:

$$\frac{C}{1000} = \frac{(\mathrm{OD}_2 - \mathrm{OD}_1)}{13,600} \times E_1 \times \frac{11}{2} \times \frac{5}{2} \times \frac{100}{\mathrm{Hb}}$$

$$C = \frac{(\mathrm{OD}_2 - \mathrm{OD}_1) \times E_1 \times 101}{\mathrm{Hb}}$$

where OD_1 is the optical density measured at 412 nm before the addition of DTNB solution, and OD_2 is the optical density after the addition of DTNB.

In the rare circumstance that the filtrate is deeply colored, $OD_2 - OD_1$ is corrected to

$$OD_2 - \frac{10}{11} OD_1$$

To obtain the more conventional expression of GSH levels as milligrams of GSH per 100 ml of red cells, 0.2 ml of whole blood may be added to 1.8 ml of distilled water and a hematocrit determination carried out on the blood sample. In this case, the concentration of glutathione (C_1) in milligrams per milliliter is obtained as follows:

$$C_1 = \frac{(OD_2 - OD_1)}{13,600} \times E_1 \times \frac{11}{2} \times \frac{0.2}{5} \times \frac{100}{Hct} \times 307$$

$$= (OD_2 - OD_1) \times \frac{310.4}{Hct} \times E_1$$

or C' (in milligram percent):

$$C' = (OD_2 - OD_1) \times \frac{31040}{Hct} \times E_1$$

where Hct is the hematocrit in percent.

The red cells of normal adults contain 6.57 ± 1.04 μmoles of GSH/g Hb (mean $\pm$ standard deviation). This is equivalent to approximately 68.5 ± 10.8 mg %.

Diminished GSH stability—that is, inability of the red cell to maintain its GSH level in the face of acetylphenylhydrazine-induced stress—is a consistent finding in glucose-6-phosphate dehydrogenase deficiency. In addition, a deficiency of gamma glutamyl cysteine synthetase or glutathione synthetase, resulting in virtual absence of red cell GSH, is a rare cause of nonspherocytic congenital hemolytic anemia. Elevated values of GSH are found in patients with myelofibrosis.

REFERENCES

1. Beutler, E., Duron, O., and Kelly, B. M.: Improved method for the determination of blood glutathione. J. Lab. Clin. Med. 61:882–890, 1963.

37

oxidized glutathione (GSSG)

A. PRINCIPLE

Oxidized glutathione (GSSG) estimations are carried out using the reaction

$$GSSG + NADPH + H^+ \xrightarrow{GR} 2GSH + NADP^+$$

The amount of NADPH oxidized bears a stoichiometric relationship to the GSSG available and is measured at 340 nm. Red cells contain several hundred times as much GSH as GSSG. Hence, oxidation of even a small proportion of red cell GSH during the preparation of the filtrate will result in large errors. This is prevented by alkylating red cell GSH by the addition of an excess of N-ethylmaleimide (NEM), and removing excess NEM by ether extraction at the same time that the TCA used for precipitation is being removed.

B. PROCEDUURE (based on Srivastava and Beutler [1])

Two and five-tenths milliliters of packed red cells are pipetted into a centrifuge tube containing 0.5 ml of 0.25 M NEM. The NEM solution and erythrocytes are mixed thoroughly and allowed to stand at 0° for 10 min. Ten microliters of the mixture are added to 10 ml of ferricyanide-cyanide solution. Two milliliters of ice-cold 30% (W/V) TCA are added to the centrifuge tube, and the contents of the tube are stirred thoroughly. After centrifugation at 1000 g for 15 min, 1.5 ml of the supernatant are extracted three times, each with about 5 ml of ice-cold ether. Excess ether is removed with a stream of nitrogen or air. The assay is carried out in the following reaction system:

	Blank (μl)	System (μl)
K$_2$HPO$_4$/KH$_2$PO$_4$, 1 M, pH 7.4, with		
4 mM EDTA	80	80
NADPH, 12 mM	10	10
TCA extract	—	900
H$_2$O	900	—

Equilibrate at 37° C or 25° C while recording absorbance
at 340 nm until rate of change of OD is constant

| Gutathione reductase, 90 U/ml | 10 | 10 |

Read the optical density for a few minutes, until the
rate of change becomes constant

A recorder expansion giving a full-scale reading of 0.1 OD unit is suitable for normal blood samples.

When using a recording spectrophotometer which does not have a provision for arbitrarily adjusting the position of the recorder pen, or when using a nonrecording spectrophotometer, the instrument should be blanked at an optical density value of 0.08, since the optical density of the system will fall.

C. ADDITIONAL BLANKS REQUIRED

None.

D. COMMENTS, CALCULATIONS, AND NORMAL VALUES

Incomplete extraction with ether will lead to a rapid decrease in the optical density of the system prior to addition of glutathione reductase, because of the instability of NADPH in an acid medium.

A straight line should be fitted to the initial rate observed and to the final rate. The distance between these lines at the time of GR addition represents the amount of NADPH oxidized, and hence the amount of GSSG in the filtrate. The concentration of GSSG in the filtrate in micromoles per gram of Hb (C) is obtained as follows:

$$C = \frac{(10/9) \times (4.24/2.49) \times (\Delta OD/6.22)}{0.01\ Hb} = \frac{\Delta OD}{Hb} \times 30.4$$

where ΔOD represents the change in optical density, and Hb is the concentration of hemoglobin in gram percent. The factor 30.4 represents the combination of the factors 10/9 (dilution in the cuvette), 4.24/2.49 (dilution while preparing the extract [see page 13]), 6.22 (the optical density of an NADPH solution containing 1 μmole/ml), and 0.01 to convert hemoglobin in gram percent to grams per milliliter. It is useful to use the fully expanded scale on the spectrophotometer recorder in making these measurements, since the amount of GSSG in red cells is very small, and one must then remember to correct the optical density units for the expansion employed.

The red cells of normal adults contain 0.0123 ± 0.0045 μmole GSSG/g Hb (mean $\pm$ standard deviation). Elevations of GSSG levels have been found to occur in G-6-PD-deficient red cells.

REFERENCE

1. Srivastava, S. K., and Beutler, E.: Accurate measurement of oxidized glutathione content of human, rabbit, and rat red blood cells. Anal. Biochem. 25:70–76, 1968.

38

pyruvate

A. PRINCIPLE

In the presence of lactate dehydrogenase, pyruvate is quantitatively reduced to lactate by NADH in the reaction

$$\text{Pyruvate} + \text{NADH} + \text{H}^+ \overset{\text{LDH}}{\rightleftharpoons} \text{Lactate} + \text{NAD}^+$$

The amount of NADH oxidized bears a stoichiometric relationship to the pyruvate present and is measured at 340 nm.

B. PROCEDURE

Add 1 ml of whole blood to 3 ml of ice-cold 4 percent PCA. Mix well and prepare a neutralized extract from 1.5 ml of the supernatant, adjusting volume to 2 ml (see pages 16 and 17).

	Blank (μl)	System (μl)
K_2HPO_4/KH_2PO_4, 1 M, pH 8.0	200	200
NADH, 2 mM	100	100
PCA extract	—	500
H_2O	690	190

Read baseline at 37° at 340 nm

LDH, 100 U/ml	10	10

Measure OD at 340 nm until no further change occurs

A recorder expansion giving a full-scale reading of 0.1 OD unit is suitable for normal blood samples. When using a recording spectrophotometer which does not have a provision for arbitrarily adjusting the position of the recorder pen, or when using a nonrecording spectrophotometer, the instrument should be blanked at an optical density value of 0.08, since the optical density of the system will fall.

C. ADDITIONAL BLANKS REQUIRED

None.

D. COMMENTS, CALCULATIONS, AND NORMAL VALUES

Although pure solutions of pyruvate are quite stable, pyruvate may disappear rapidly from perchloric acid filtrates of blood. Accordingly, extracts prepared for pyruvate estimation should not be stored and should be neutralized immediately prior to assay.

Since pyruvate diffuses through the red cell membrane and is distributed through plasma and red cells, pyruvate concentrations of blood (C) are expressed as micromolar:

$$C = \frac{\Delta OD \times 10.13 \times 1000}{6.22} = \Delta OD \times 1628$$

The factor 10.13 is to account for the 1:3.8 dilution of blood in the original extract (assuming 80% water content for blood [see page 13]), the 1.5:2.0 dilution during neutralization, and the 1:2 dilution in the cuvette. The results are expressed as micromolar; therefore, it is necessary to multiply by 1000, and 6.22 is the millimolar extinction coefficient of NADH.

The normal concentration of pyruvate in freshly drawn blood is 53.3 ± 21.5 μM (mean ± standard deviation). Estimation of pyruvate levels is of interest chiefly in studies of erythrocyte glycolysis.

39

lactate

In the presence of NAD and lactic dehydrogenase, lactate is oxidized to pyruvate in the reaction

$$NAD^+ + \text{lactate} \underset{}{\overset{LDH}{\rightleftharpoons}} NADH + H^+ + \text{pyruvate}$$

A. PRINCIPLE

At physiologic pH levels, the lactate-pyruvate equilibrium lies far in the direction of lactate. However, by the use of a high pH, a high concentration of NAD, and hydrazine to remove pyruvate formed from the equilibrium, it is possible to push the reaction to the right. Under these conditions, lactate stoichiometrically reduces NAD to NADH, and the quantity of lactate in the mixture may be measured from the change in optical density at 340 nm.

B. PROCEDURE

Add 1 ml of whole blood to 3 ml of ice-cold 4% PCA. Mix well and prepare a neutralized extract from 1.5 ml of the supernatant, adjusting the volume to 2 ml (see pages 16 and 17).

	Blank (μl)	System (μl)
Glycine hydrazine buffer*	500	500
NAD, 5 mM	200	200
PCA extract	—	100
H$_2$O	290	190

Read baseline at 37° at 340 nm

LDH, 2500 U/ml	10	10

Measure OD at 340 nm until no further change occurs

* Stock solution contains 7.5 g glycine, 5.2 g hydrazine sulfate, and 0.2 g disodium EDTA in 49 ml. Store in refrigerator. Before use, add 5.1 ml of 2 M NaOH to 4.9 ml of stock solution.

The recorder expansion giving a full-scale reading of 0.4 OD unit is suitable for normal blood samples.

C. BLANKS REQUIRED

None.

D. COMMENTS, CALCULATIONS, AND NORMAL VALUES

Special care is required to avoid contaminating glassware used for lactate determinations with sweat from the hands, since sweat is very rich in lactate.

Since lactate diffuses through the red cell membrane and is distributed through plasma and red cells, lactate concentrations of blood (C) are expressed as micromolar:

$$C = \frac{\Delta OD \times 50.67 \times 1000}{6.22} = \Delta OD \times 8146$$

The factor is to account for the 1:3.8 dilution of blood in the original filtrate (assuming 80% water content for blood [see page 13]), the 1.5:2.0 dilution during neutralization, and the 1:10 dilution in the cuvette. The results are expressed as micromolar; therefore, it is necessary to multiply by 1000, and 6.22 is the millimolar extinction coefficient of NADH.

The normal concentration of lactate in freshly drawn blood is 932 ± 211 μM (mean $\pm$ standard deviation). Estimation of lactate levels are of interest chiefly in studies of erythrocyte glycolysis.

40

glucose-6-P and fructose-6-P

A. PRINCIPLE

When glucose-6-phosphate dehydrogenase (G-6-PD) is added to a solution containing glucose-6-P and NADP, the NADP is reduced in the reaction

$$\text{Glucose-6-P} + \text{NADP}^+ \xrightarrow{\text{G-6-PD}} \text{6-PGA} + \text{NADPH} + \text{H}^+$$

The increase in fluorescence occurring when NADP is reduced to NADPH is recorded and represents the glucose-6-P content of the sample.

Next, glucose phosphate isomerase (GPI) is added to the same mixture. This results in the conversion of fructose-6-P in the mixture to glucose-6-P, which is oxidized by G-6-PD reducing additional NADP to NADPH. The change in fluorescence occurring after glucose phosphate isomerase (GPI) is added is a measure of the fructose-6-P concentration. The change of fluorescence occurring after the addition of aliquots of a standard glucose-6-phosphate solution makes possible calculation of the results.

B. PROCEDURE

A perchloric acid extract is prepared by adding a 2-ml blood sample to 8 ml of ice-cold 4% perchloric acid. If levels representing the in vivo concentrations are desired, blood should be added directly from the syringe, without cooling or any other operation delaying transfer to the PCA. Seven milliliters of the supernatant after centrifugation are neutralized with 1 M K_2CO_3 and the volume adjusted to 10 ml (see pages 16-17). The neutralized extract is centrifuged for 10 min at 12,000 g at 4° C to remove particulates, and the supernatant is used as indicated below. The mixed reagents should be equilibrated at 37° and readings made in a temperature-controlled filter fluorometer, such as a Turner Model 111, with a Corning 7-60 primary and 3-72 secondary filter.

	μl
Tris-HCl, 1 M, EDTA, 5 mM, pH 8.0	500
NADP, 2 mM	1000
β-Mercaptoethanol, 14 M	5
Extract	1200
H_2O	1300

Obtain a baseline

G6PD, ~150 U/ml	5

Record fluorescence until a stable reading is obtained

GPI, ~1500 U/ml	5

Record fluorescence until a stable reading is obtained

Glucose-6-P standard,* 50 μM	10

Record fluorescence until a stable reading is obtained

Glucose-6-P standard,* 50 μM	20

Record fluorescence until a stable reading is obtained

* See Section D.

C. BLANKS REQUIRED

A blank in which water is substituted for the perchloric acid extract should be carried through the same series of estimations.

D. COMMENTS, CALCULATIONS, AND NORMAL VALUES

Since the amount of glucose-6-P and fructose-6-P being determined is very small, a sensitive filter fluorometer is required. It is important that all solutions be free of lint and other particulates, and filtration of reagents through hard filter paper may be helpful in this respect. Even under the best circumstances, a recording will show irregularities, and measurements of changes in fluorescence should be based on the average fluorescence reading before and after each addition of enzyme or substrate. The concentration of glucose-6-P in the standard should be assayed spectrophotometrically. This can be done using the reaction mixture for G-6-PD (WHO-37°) assay (page 67). Ten microliters of G-6-PD ($\sim$10 U/ml) should be substituted for hemolyzate and a 1.0 mM (by weight) solution of glucose-6-P for the 6 mM solution specified. The optical density of the mixture, lacking only G-6-PD, should be recorded at 340 nm and the optical density increase after the addition of G-6-PD should be measured. The expected change is 0.622 OD unit. The ratio of the observed change to the expected change is the millimolarity of the glucose-6-P solution. The solution should then be diluted appropriately to give a concentration of 50 μM. C, the concentration of fructose-6-P or glucose-6-P in micromoles per liter of red cells, is given by

$$
\begin{aligned}
C &= \frac{\Delta F_s}{\Delta F_{st}} \times 0.125 \times \frac{4}{1.2} \times \frac{10}{7} \times \frac{9.6}{2} \times \frac{1}{\text{Hct}} \\
&= \frac{\Delta F_s \times 2.86}{\Delta F_{st} \times \text{Hct}}
\end{aligned}
$$

where ΔF_s is the change of fluorescence produced by glucose-6-P or fructose-6-P, Hct is the hematocrit (a hematocrit of 50% = 0.50), and ΔF_{st} is the change of fluorescence produced by 10 μl of the 50 μM standard. This value is best obtained by averaging the change produced by 10 μl of the standard and half of the change produced by 20 μl of the standard. If 20 μl of the standard does not produce approximately twice the change produced by 10 μl, the assay is invalid. The concentration of the standard in the cuvette is 0.125 μM. The fractions 4/1.2, 10/7, and 9.6/2 represent, respectively, the dilution of the sample in the cuvette, the dilution of the perchloric acid during and after neutralization, and the dilution of the red cells in perchloric acid (see page 13).

The normal concentration of glucose-6-P in red cells is 27.8 $\pm$ 7.5

nmoles/ml RBC (mean ± standard deviation). The normal concentration of fructose-6-P is 9.3 ± 2.0 nmoles/ml RBC (mean ± standard deviation). In patients with high reticulocyte counts, the levels of glucose-6-P and fructose-6-P are increased, but the normal 3:1 ratio of glucose-6-P to fructose-6-P is maintained except in patients with glucose phosphate isomerase deficiency, in whom the ratio of G-6-P to F-6-P increases.

41

glyceraldehyde-3-P (GAP), dihydroxyacetone-P (DHAP), and fructose-diP (FDP)

A. PRINCIPLE

Glyceraldehyde-3-phosphate is determined by measuring the increase in fluorescence as NAD is reduced to NADH in the reaction

$$\text{GAP} + \text{NAD}^+ \xrightarrow[\text{arsenate}]{\text{GAPD}} 3\text{PGA} + \text{NADH} + \text{H}^+$$

Arsenate is used in this reaction to substitute for phosphate, the physiologic substrate in the glyceraldehyde-3-phosphate dehydrogenase reaction. The arsenic acid anhydride formed undergoes spontaneous arsenolysis, and the reaction is therefore irreversible.

When triose phosphate isomerase (TPI) is added, dihydroxyacetone phosphate (DHAP) is converted to GAP in the isomerization reaction

$$\text{GAP} \underset{}{\overset{\text{TPI}}{\rightleftharpoons}} \text{DHAP}$$

The GAP formed reduces additional NAD to NADH, so that the increase in optical density after the addition of TPI represents the amount of DHAP in the extract.

Aldolase cleaves fructose diphosphate (FDP) to GAP and DHAP in the reaction

$$\text{FDP} \xrightleftharpoons{\text{aldolase}} \text{DHAP} + \text{GAP}$$

The DHAP and GAP formed serve as further substrate for the reduction of NAD in the GAPD reaction. Thus the increase in fluorescence after addition of aldolase represents the amount of fructose diphosphate. Finally, a fructose diphosphate standard is added to permit calculation of the results.

B. PROCEDURE

The same red cell extract described in Chapter 40 may be used. The mixed reagents should be equilibrated at 37° and readings made in a temperature-controlled filter fluorometer, such as a Turner Model 111, with a Corning 7-60 primary filter and 3-72 secondary filter.

	μl
Tris-HCl, 1 M, EDTA, 5 mM, pH 8.0	400
NAD, 10 mM	800
Sodium arsenate, pH 8.0, 300 mM	600
β-Mercaptoethanol, 14 M	5
Extract	2200

Obtain a baseline

GAPD, ~550 U/ml   10

Record fluorescence until a stable reading is obtained

TPI, 24,000 U/ml   5

Record fluorescence until a stable reading is obtained

Aldolase, ~100 U/ml   10

Record fluorescence until a stable reading is obtained

FDP standard,* 50 μM 10

Record fluorescence until a stable reading is obtained

FDP standard,* 50 μM 20

Record fluorescence until a stable reading is obtained

* See Section D.

C. BLANKS REQUIRED

A blank in which water is substituted for the perchloric acid extract should be carried through the same series of estimations.

D. COMMENTS, CALCULATIONS, AND NORMAL VALUES

Since the amount of GAP, DHAP, and FDP being determined is very small, a sensitive filter fluorometer is required. It is important that all solutions be free of lint and other particulates, and filtration of reagents through hard filter paper may be helpful in this respect. Even under the best circumstances, the recording will show irregularities, and measurements of changes in fluorescence should be based on the average fluorescence reading before and after each addition of enzyme or substrate once the endpoint has been reached. The concentration of fructose diphosphate in the standard should be assayed spectrophotometrically. This can be done using the reaction mixture for aldolase assay (page 46). Ten microliters of aldolase (approximately 100 U/ml) should be substituted for hemolyzate. A 0.5 mM (by weight) solution of FDP is substituted for the 0.01 M solution specified in the aldolase assay. The optical density of a mixture lacking only aldolase, should be recorded at 340 nm and the optical density change after the addition of aldolase measured. The expected change is 0.622 OD unit. The ratio of the observed change to the expected change times 0.5 is the millimolarity of the FDP solution. The solution should then be diluted appropriately to give a concentration of 50 μM. C, the concentration of DHAP or GAP in micromoles per liter of red cells, is given by

$$
C = \frac{\Delta F_s}{\Delta F_{st}} \times 0.25 \times \frac{4}{2.2} \times \frac{10}{7} \times \frac{9.6}{2} \times \frac{1}{\text{Hct}}
$$

$$
= \frac{\Delta F_s \times 3.12}{\Delta F_{st} \times \text{Hct}}
$$

where ΔF_s is the change of fluorescence produced by DHAP or GAP, Hct is the hematocrit (a hematocrit of 50% = 0.50), and ΔF_{st} is the change of fluorescence produced by 10 μl of the 50 μM standard. This value is best obtained by averaging the change produced by 10 μl of the standard and half of the change produced by 20 μl of the standard. If 20 μl of the standard do not produce approximately twice the change produced by 10 μl, the assay is invalid. The concentration of the standard in the cuvette is represented by 0.250 since the 0.125 μM FDP will reduce 0.250 μM NAD to NADH. The fractions 4/2.2, 10/7, and 9.6/2 represent, respectively, the dilution of the sample in the cuvette, the dilution of the perchloric acid during and after neutralization, and the dilution of the red cells in perchloric acid (see page 13).

The FDP concentration of the erythrocytes in micromoles per liter is given by

$$
C = \frac{\Delta F_s \times 1.56}{\Delta F_{st} \times \text{Hct}}
$$

because each mole of FDP results in the reduction of 2 moles of NAD to NADH.

Normal red cells do not contain detectable amounts of GAP. The normal concentration of DHAP is 9.4 ± 2.8 nmoles/ml RBC (mean $\pm$ standard deviation). The normal concentration of FDP is 1.9 ± 0.6 nmoles/ml RBC (mean $\pm$ standard deviation). In patients with high reticulocyte counts, the levels of DHAP and FDP are increased. The levels of FDP rise very rapidly after blood is drawn, especially if the sample is cooled.

42

3-phosphoglyceric acid (3-PGA)

A. PRINCIPLE

3-Phosphoglyceric acid (3-PGA) is phosphorylated to 1,3-diphosphoglyceric acid (1,3-DPG) in the phosphoglycerate kinase (PGK) reaction:

$$3\text{-PGA} + \text{ATP} \underset{\text{Mg}^{2+}}{\overset{\text{PGK}}{\rightleftharpoons}} 1,3\text{-DPG} + \text{ADP}$$

The 1,3-DPG formed is reduced and dephosphorylated in the reverse glyceraldehyde phosphate dehydrogenase (GAPD) reaction:

$$1,3\text{-DPG} + \text{NADH} + \text{H}^+ \overset{\text{GAPD}}{\rightleftharpoons} \text{GAP} + \text{P}_i + \text{NAD}^+$$

Finally, a 3-PGA standard is added to permit calculation of the results.

B. PROCEDURE

The same red cell extract described in Chapter 40 may be used. A reaction mixture is prepared and measurements are made as indicated below. The mixed reagents should be equilibrated at 37° and readings made in a temperature-controlled filter fluorometer, such as a Turner Model 111, with a Corning 7-60 primary and 3-72 secondary filter.

	μl
Potassium phosphate buffer, pH 7.4, 100 mM	500
ATP, 10 mM	400
MgCl$_2$, 100 mM	40
β-Mercaptoethanol, 14 M	5
NADH, 0.2 mM	10
GAPD, 550 U/ml	5
Extract	500
H$_2$O	2540

Record fluorescence until a constant, slow rate is obtained

PGK, 3200 U/ml	5

Record fluorescence until a constant, slow rate is obtained

3-PGA standard,* 50 μM	10

Record fluorescence until a constant, slow rate is obtained

3-PGA standard,* 50 μM	20

Record fluorescence until a constant, slow rate is obtained

* See Section D.

C. BLANKS REQUIRED

A blank in which water is substituted for the perchloric acid extract should be carried through the same series of estimations.

D. COMMENTS, CALCULATIONS, AND NORMAL VALUES

Since the amount of 3-PGA being determined is very small, a sensitive filter fluorometer is required. It is important that all solutions be free of lint and other particulates, and filtration of reagents through hard filter paper may be helpful in this respect. Even under the best circumstances, the recording will show irregularities, and measurements of changes in fluorescence should be based on the average fluorescence reading before and after each addition of enzyme or substrate once the endpoint has been reached. The 3-PGA standard should be assayed spectrophotometrically. This may be done by using the reaction mixture for phosphoglycerate kinase assay (page 53) substituting, however, 5μl of PGK (~3200 U/ml) for the hemolyzate, and a 1 mM (by weight) solution of 3-PGA for the 0.01 M solution specified. The optical density of the mixture, lacking only PGK, should be recorded, and the fall in optical density after the addition of PGK measured. The expected change is 0.622 OD unit. The ratio of the change found to the expected change is the millimolarity of the 3-PGA solution, and it should be diluted appropriately to provide a concentration of 50 μM. C, the concentration of 3-PGA in micromoles per liter of red cells, is given by

$$C = \frac{\Delta F_s}{\Delta F_{st}} \times 0.125 \times \frac{4}{0.5} \times \frac{10}{7} \times \frac{9.6}{2} \times \frac{1}{Hct}$$
$$= \frac{\Delta F_s \times 6.86}{\Delta F_{st} \times Hct}$$

where ΔF_s is the change of fluorescence produced by 3-PGA, Hct is the hematocrit (a hematocrit of 50% = 0.50), and ΔF_{st} is the change of fluorescence produced by 10 μl of the 50 μM standard. This value is best obtained by averaging the change produced by 10 μl of the standard and half of the change produced by 20 μl of the standard. If 20 μl of the standard do not produce approximately twice the change produced by 10 μl, the assay is invalid. The concentration of the standard in the cuvette is 0.125 μM. The fractions 4/0.5, 10/7, and 9.6/2 represent, respectively, the dilution of the sample in the cuvette, the dilution of the perchloric acid during and after neutralization, and the dilution of the red cells in perchloric acid (see page 13).

The normal concentration of 3-PGA in red cells is 44.9 ± 5.1 nmoles/ml RBC (mean ± standard deviation).

43

2-phosphoglyceric acid (2-PGA) and phosphoenolpyruvate (PEP)

A. PRINCIPLE

The estimation of 2-phosphoglyceric acid (2-PGA) and phosphoenolpyruvate (PEP) depends on their sequential conversion to pyruvate, which is reduced to lactate by NADH through the lactate dehydrogenase reaction. First, however, endogenous pyruvate must be removed. This is accomplished by addition of sufficient NADH to reduce the pyruvate present in the extract to lactate. After this has been done, phosphoenolpyruvate is converted to pyruvate in the pyruvate kinase (PK) reaction:

$$PEP + ADP \xrightarrow[\text{Mg}^{2+}]{\text{PK}} \text{Pyruvate} + ATP$$

The pyruvate is reduced in the lactate dehydrogenase (LDH) reaction:

$$\text{Pyruvate} + NADH + H^+ \overset{\text{LDH}}{\rightleftharpoons} \text{Lactate} + NAD^+$$

The decrease in fluorescence resulting from the oxidation of NADH is measured and represents the amount of PEP in the extract.

Next, enolase is added to convert 2-PGA in the mixture to PEP:

$$\text{2-PGA} \overset{\text{enolase}}{\underset{\longleftarrow}{\rightleftharpoons}} \text{PEP}$$

The PEP formed is converted to pyruvate and finally to lactate in the reaction shown above. The decrease in fluorescence which occurs after the addition of enolase represents the 2-PGA in the mixture. Finally, 2-PGA standards are added to permit calculation of the results.

B. PROCEDURE

The perchloric acid extract described on page 16 may be used. A reaction mixture is prepared and measurements are made as indicated below. The mixed reagents should be equilibrated at 37° and readings made in a temperature-controlled filter fluorometer, such as Turner Model 111, with a Corning 7-60 primary and 3-72 secondary filter.

	μl
Potassium phosphate buffer, pH 7.4, 0.5 M	500
ADP, 10 mM	400
MgCl$_2$, 0.1 M	40
NADH, 0.2 M	160
Extract	1500
H$_2$O	1400
LDH, 150 U/ml	5

After the LDH reaction is complete, as indicated by stabilization of fluorescence (approximately 5 min), add additional 10-μl aliquots of NADH until addition of NADH causes an increase of fluorescence which is no longer followed by any substantial decrease of fluorescence.

Record fluorescence until the baseline is stable (approximately 15 min)

PK, ~3700 U/ml   5

Record fluorescence until a constant, slow rate is obtained

Enolase, ~410 U/ml   10

Record fluorescence until a constant, slow rate is obtained

2-PGA standard,* 50 μM 10

Record fluorescence until a constant, slow rate is obtained

2-PGA standard,* 50 μM 20

Record fluorescence until a constant, slow rate is obtained

* See Section D.

C. BLANKS REQUIRED

A blank in which water is substituted for the perchloric acid should be carried through the same series of estimations.

D. COMMENTS, CALCULATIONS, AND NORMAL VALUES

Since the amount of 2-PGA and PEP being determined is very small, a sensitive filter fluorometer is required. It is important that all solutions be free of lint and other particulates, and filtration of reagents through hard filter paper may be helpful in this respect. Even under the best circumstances, the recording will show irregularities, and measurements of changes in fluorescence should be based on the average fluorescence reading before and after each addition of enzyme or substrate once the endpoint has been reached. The concentration of 2-PGA in the standard should be assayed spectrophotometrically. This can be done using the reaction mixture for enolase (page 59). Ten microliters of enolase (~10 U/ml) should be substituted for hemolyzate, and a 1.0 mM (by weight) solution of 2-PGA substituted for the 10 mM solution specified. The optical density of a mixture lacking only enolase, should be recorded at 340 nm and the optical density decrease after the addition of enolase measured. The expected change is 0.622 OD unit. The ratio of the observed change to the expected change is the millimolarity of the 2-PGA solution. The solution should then be diluted appropriately to give a concentration of 50 μM. C, the concentration of 2-PGA and PEP in micromoles per liter of red cells, is given by

$$C = \frac{\Delta F_s}{\Delta F_{st}} \times 0.125 \times \frac{4}{1.5} \times \frac{10}{7} \times \frac{9.6}{2} \times \frac{1}{Hct}$$

$$= \frac{\Delta F_s \times 2.29}{\Delta F_{st} \times Hct}$$

where ΔF_s is the change of fluorescence produced by PEP or 2-PGA, Hct is the hematocrit (a hematocrit of 50% = 0.50), and ΔF_{st} is the change of fluorescence produced by 10 μl of the 50 μM standard. This value is best obtained by averaging the change produced by 10 μl of the standard and half of the change produced by 20 μl of the standard. If 20 μl of the standard do not produce approximately twice the change produced by 10 μl, the assay is invalid. The concentration of the standard in the cuvette is 0.125 μM. The fractions 4/1.5, 10/7, and 9.6/2 represent, respectively, the dilution of the sample in the cuvette, the dilution of the perchloric acid during and after neutralization, and the dilution of the red cells in perchloric acid (see page 13).

The normal concentration of 2-PGA in red cells is 7.3 ± 2.5 nmoles/ml RBC (mean ± standard deviation). The normal concentration of PEP is 12.2 ± 2.2 nmoles/ml RBC (mean ± standard deviation). Elevated PEP and 2-PGA levels are found in pyruvate kinase deficiency.

part VI

screening techniques

44

general principles

The methods presented in Parts III and IV permit the quantitative estimation of red cell enzyme activities. In practice, however, it is often enough to know whether or not there is a marked deficiency of the activity of the enzyme in question. Slight deviations from normal are usually not likely to be of clinical importance. For this reason, a number of screening techniques for the detection of enzyme deficiencies have been developed. These techniques are useful for the laboratory that does not have the instrumentation required for quantitative enzyme assays, or in any circumstance in which it is merely sufficient to ascertain whether or not a severe deficiency state exists. All of the procedures which are presented in this section depend on pyridine nucleotide-linked reactions. Instead of measuring the rate of oxidation or reduction of a pyridine nucleotide spectrophotometrically, fluorescence visible to the naked eye is used as an indicator. Reduced pyridine nucleotides fluoresce when illuminated with long-wave ultraviolet light, while no such fluorescence occurs with oxidized pyridine nucleotides.

In each screening test, a sample of blood is added to the screening solution and spotted on Whatman No. 1 or any other conveniently available nonfluorescing filter paper. After drying, the spots may be examined under long-wave ultraviolet light using an ordinary hand lamp.

45

glucose phosphate isomerase screening test

A. PRINCIPLE

The principle of the test is identical to that used in the spectrophotometric assay, but visual observation of the development of fluorescence in the reaction mixture as NADP is reduced to NADPH serves as an indicator of the enzyme activity rather than spectrophotometric measurement of the change in optical density at 340 nm.

B. SCREENING MIXTURE

Reagent	Volume (μl)
Tris-HCl, 1 M, EDTA, 5 mM, pH 8.0	100
MgCl$_2$, 0.1 M	100
NADP, 2 mM	200
G-6-PD, 10 U/ml (Torula yeast, Sigma)	100
H$_2$O	400
Fructose-6-P, 20 mM in 10 mM KH$_2$PO$_4$	100

All of the components of the reaction mixture except for fructose-6-P may be premixed and are stable frozen for at least 4 months. The fructose-6-P may also be stored frozen, and should be added to the reaction mixture on the day of use.

C. PROCEDURE (based on Blume and Beutler [1])

Five microliters of blood are added to 100 μl of water. Five microliters of the lyzate are mixed with 100 μl of the reaction mixture and allowed to stand at room temperature. A control tube, containing only reaction mixture, is also allowed to stand at room temperature. A spot is made on filter paper immediately, at 15 min, and after 30 min. The spots are allowed to dry and then examined under long-wave ultraviolet light.

D. INTERPRETATION

The first spot should have little or no fluorescence in each case. Normal samples will fluoresce strongly at 15 and at 30 min. Blood from patients with severe glucose phosphate isomerase deficiency will show little or no more fluorescence than the spot made from the reaction mixture without hemolyzate. This control will fluoresce slightly, because commercial fructose-6-P preparations are invariably contaminated with some glucose-6-P which will result in the reduction of some NADP to NADPH.

It is useful always to examine a normal control sample side by side with the patient's samples.

REFERENCE

1. Blume, K., and Beutler, E. Detection of glucose-phosphate isomerase deficiency by a screening procedure. Blood 39:685–687, 1972.

46

triose phosphate isomerase screening test

A. PRINCIPLE

The principle of the procedure is identical to that used in the assay for triose phosphate isomerase (see Chapter 11), but the loss of fluorescence of NADH under ultraviolet light serves as an indicator of enzyme activity rather than spectrophotometric measurement of the change of absorbance.

B. SCREENING MIXTURE

Each milliliter of screening mixture contains the reagents indicated below. When preparing the mixture, it is important that the buffer is added to the mixture first, so that the NADH will not be destroyed by the acid glyceraldehyde-3-phosphate.

Reagent	Volume (μl)
Triethanolamine buffer, 0.1 M, pH 8.0	500
NADH, 7 mM	100
DL-Glyceraldehyde-3-P, 50 mg/ml ($\sim$290 mM)	10
α-Glycerophosphate dehydrogenase, 8 U/ml	10
EDTA, 0.027 M	200
H_2O	180

C. PROCEDURE (based on Kaplan et al. [1])

Whole blood samples collected in EDTA, heparin, or ACD solution are satisfactory. Ten microliters of whole blood are added to 3 ml of distilled water and mixed well. Ten microliters of the hemolyzate are added to 100 μl of screening mixture, and the mixture is spotted on filter paper. The remainder is permitted to stand for 20 or 30 min, and a second spot is made. The spots are allowed to dry for a few minutes and are examined under ultraviolet light.

D. INTERPRETATION

The first spot should fluoresce brightly. The second spot will normally show little or no fluorescence. In TPI deficiency, however, it will fluoresce nearly as brightly as the spot made before incubation of the mixture. It is useful always to examine a normal control sample side by side with the patient's sample.

REFERENCE

1. Kaplan, J. C., Shore, N., and Beutler, E.: The rapid detection of triose phosphate isomerase deficiency. Amer. J. Clin. Pathol. 50:656–658, 1968.

47

pyruvate kinase screening test

A. PRINCIPLE

The principle of the screening procedure is identical to that used in the assay for pyruvate kinase (see Chapter 17), but loss of fluorescence of NADH under long-wave ultraviolet light is observed rather than measuring the change of absorbance spectrophotometrically. Since leukocytes of patients with pyruvate kinase deficiency have normal pyruvate kinase activity, it is particularly important to exclude as many white cells as possible from the sample being screened. This is accomplished by two means. First of all, the blood is centrifuged and the buffy coat is removed. Second, hypotonic lysis is employed, rather than a lytic agent such as digitonin or saponin. In this way, most of the leukocytes remain intact and their enzyme is not released into the mixture.

B. SCREENING MIXTURE

Reagent	Volume (μl)
Phosphoenolpyruvate (PEP), *cyclohexyl ammonium salt*, 0.15 M (neut.)	30
ADP, 30 mM (neut.)	100
NADH, 15 mM (neut.)	100
MgCl$_2$, 80 mM	100
K$_2$HPO$_4$/KH$_2$PO$_4$ buffer, 0.25 M, pH 7.4	50
H$_2$O	620

The reaction mixture is stable for only about 4 days when frozen. This is because of the relative instability of NADH. It is possible to prepare a partial reaction mixture by substituting water for the NADH solution. The frozen partial mixture is stable indefinitely. When needed, sufficient partial mixture is added to a preweighed dry NADH vial (Sigma) to give a concentration of 1 mg/ml. Thus, 0.5 ml of the partial mixture may be added to a 0.5 mg NADH vial.

138

C. PROCEDURE (based on Beutler [1])

Blood collected in heparin, EDTA, or ACD is satisfactory. The blood sample is centrifuged, and the plasma and buffy coat are carefully removed by aspiration. A 20% suspension of red cells is prepared by adding four volumes of 0.9% sodium chloride solution. Ten microliters of cell suspension are added to 100 μl of the screening mixture. A portion of the mixture is spotted on filter paper immediately, and the remainder is incubated at 37° for 30 min. After incubation, a second spot is made. After drying on the paper, the spots are examined under long-wave ultraviolet light.

D. INTERPRETATION

The first spot should fluoresce brightly, but fluorescence has disappeared from the second spot when normal blood is tested. A pyruvate kinase-deficient sample shows fluorescence in both spots. By making spots every 10 min, it is possible to identify provisionally heterozygotes for PK deficiency. It is useful always to examine a normal control sample side by side with the patient's sample.

REFERENCE

1. Beutler, E.: A series of new screening procedures for pyruvate kinase deficiency, glucose-6-phosphate dehydrogenase deficiency, and glutathione reductase deficiency. Blood 28:553–562, 1966.

48

glucose-6-phosphate dehydrogenase screening test

A. PRINCIPLE

In the presence of glucose-6-phosphate dehydrogenase and NADP, glucose-6-phosphate is oxidized to 6-phosphogluconate in the reaction

$$\text{Glucose-6-P} + \text{NADP}^+ \xrightarrow{\text{G-6-PD}} \text{6-Phosphogluconate} + \text{NADPH} + \text{H}^+$$

Since phosphogluconate dehydrogenase (6-PGD) is present in virtually all hemolyzates, further reduction of NADP occurs in the reaction

$$\text{6-Phosphogluconate} + \text{NADP}^+ \xrightarrow{\text{6-PGD}} \text{Ribulose-5-P} + \text{NADPH} + \text{H}^+$$

When mildly G-6-PD-deficient hemolyzates are incubated with glucose-6-P and NADP, a small amount of NADPH is formed. In the presence of GSSG, provided in the screening mixture, this is reoxidized in the glutathione reductase reaction:

$$\text{GSSG} + \text{NADPH} + \text{H}^+ \xrightarrow{\text{GR}} 2\text{GSH} + \text{NADP}^+$$

Thus, the screening test measures, in effect, the difference between approximately twice the glucose-6-phosphate dehydrogenase activity and the glutathione reductase activity.

B. SCREENING MIXTURE

One milliliter of screening mixture contains the following:

Reagent	Volume (μl)
Glucose-6-P, 0.01 M	100
NADP, 7.5 mM	100
Saponin (Sigma), 1%	200
Tris-HCl buffer, 750 mM, pH 7.8	300
GSSG, 8 mM	100
H$_2$O	200

The mixture is stable, in the frozen state, for several months. A similar reagent may also be purchased, in lyophilized form, from Hyland Laboratories, 3300 Hyland Avenue, Costa Mesa, Calif. 92626.

C. PROCEDURE (based on Beutler and Mitchell [1])

Blood collected in heparin, EDTA, or ACD solution is satisfactory. Blood which is several weeks old and even spots of blood collected on filter paper and dried may be used. Ten microliters of blood are added to 100 μl of screening mixture, and a spot is made on filter paper. The mixture is allowed to incubate at room temperature for 5 to 10 min, and a second spot is made.

D. INTERPRETATION

In normal samples, the first spot may fluoresce slightly and the second spot will fluoresce brightly. Deficient samples will show little or no fluorescence in either spot. It is useful always to examine a normal control sample side by side with the patient's sample.

REFERENCE

1. Beutler, E., and Mitchell, M.: Special modifications of the fluorescent screening method for glucose-6-phosphate dehydrogenase deficiency. Blood 32:816–818, 1968.

49

glutathione reductase screening test

A. PRINCIPLE

The principle of the test is identical to that used in the spectrophotometric assay, but the loss of fluorescence of the reaction mixture as NADPH is oxidized serves as an indicator of enzyme activity rather than spectrophotometric measurement of the change in optical density at 340 nm.

B. SCREENING MIXTURE

Reagent	Volume (μl)
K_2HPO_4/KH_2PO_4 buffer, 250 mM, pH 7.4	600
Saponin (Sigma), 1%	200
GSSG, 33 mM	100
NADPH, 15 mM	100

The mixture is stable for about 10 days in the frozen state. A partial reaction mixture may be prepared substituting water for NADPH. This mixture is stable indefinitely in the frozen state. It may be mixed with a preweighed vial of NADPH (Sigma), adding sufficient partial reaction mixture to give a final NADPH concentration of 1 mg/ml.

C. PROCEDURE (based on Beutler [1])

Blood collected in heparin, ACD, or EDTA may be used. Blood stored for at least 3 weeks is satisfactory. Ten microliters of blood are added to 100 μl of screening mixture. A spot is made on filter paper immediately and at 15-min intervals during incubation at 37°. The spots are allowed to dry and are examined under long-wave ultraviolet light.

D. INTERPRETATION

The first spot should fluoresce in every case. If the activity of glutathione reductase is high, fluorescence will disappear within the first half-hour of incubation. Samples with decreased glutathione reductase activity may show continued fluorescence for an hour or even longer. It is useful always to examine a normal control sample side by side with the patient's sample.

REFERENCE

1. Beutler, E.: A series of new screening procedures for pyruvate kinase deficiency, glucose-6-phosphate dehydrogenase deficiency, and glutathione reductase deficiency. Blood 28:553–562, 1966.

50

NADH methemoglobin reductase (NADH diaphorase) screening test

A. PRINCIPLE

NADH methemoglobin reductase (NADH diaphorase) (MR) catalyzes the reduction of methemoglobin under physiologic circumstances. However, the dye dichlorophenol-endophenol (DCIP) also serves as a substrate for this enzyme, being reduced to the colorless leukoform (DCIPH$_2$). The screening procedure depends on the reaction

$$\text{NADH} + \text{H}^+ + \text{DCIP} \xrightarrow{\text{MR}} \text{NAD}^+ + \text{DCIPH}_2$$

The oxidation of NADH to NAD can be observed visually through the loss of its fluorescence under long-wave ultraviolet light. The blood sample must first be treated with nitrite to oxidize hemoglobin to methemoglobin, since hemoglobin will reduce DCIP nonenzymatically.

B. SCREENING MIXTURE

One milliliter of screening mixture may be conveniently prepared as follows, from stable reagents:

Reagents	Volume (μl)
Tris-HCl, 0.06 M, pH 7.6, containing 2.7 mM Na-EDTA	1000
DCIP, 19 mM (6.25 mg/ml)	10
Saponin (Sigma), 1%	200

Add this mixture to a preweighed vial containing 0.5 mg of dry NADH (e.g., from Sigma Chemical Co.)

C. PROCEDURE (based on Kaplan et al. [1])

Six microliters of freshly prepared 0.18 M sodium nitrite solution are added to 0.1 ml of whole blood collected in either heparin or ACD. The mixture is allowed to stand at room temperature for 30 min, and 20 μl are added to 0.4 ml of the screening mixture. A spot is made on filter paper immediately and at 15-min intervals for 45 min.

D. INTERPRETATION

The first spot should fluoresce brightly, but fluorescence has disappeared from the spot after approximately 30 min incubation when normal blood is tested. An NADH methemoglobin reductase-deficient sample will show persistent fluorescence for 45 min and longer. It is useful always to examine a normal control sample side by side with the patient's sample.

REFERENCE

1. Kaplan, J. C., Nicolas, A.-M., Hanzlickova-Leroux, A., and Beutler, E.: A simple spot screening test for fast detection of red cell NADH-diaphorase deficiency. Blood 36:330–333, 1970.

appendix 1

activities of red cell enzymes in normal adults

Enzyme	Activity at 37° C (mean ± standard deviation)
Acetylcholinesterase	36.93 ± 3.83 IU/g Hb
Adenosine deaminase	1.11 ± 0.23 IU/g Hb
Adenylate kinase	258 ± 29.3 IU/g Hb
low S	38.0 ± 0.67%
Aldolase	3.19 ± 0.86 IU/g Hb
low S	64.1 ± 7.40%
Catalase	153117 ± 23904 IU/g Hb
Diphosphoglyceromutase	4.78 ± 0.65 IU/g Hb
Enolase	5.39 ± 0.83 IU/g Hb
low S	63.1 ± 9.27%
Epimerase	0.231 ± 0.061 IU/g Hb
Galactokinase	0.291 ± 0.006 IU/g Hb
Glucose phosphate isomerase	60.8 ± 11.0 IU/g Hb
low S	46.2 ± 2.41%
Glucose-6-phosphate dehydrogenase	8.34 ± 1.59 IU/g Hb
low S	67.1 ± 6.53%
WHO method	12.1 ± 2.09 IU/g Hb
low S	62.2 ± 4.21%
Glutathione peroxidase*	30.82 ± 4.73 IU/g Hb
Glutathione reductase without FAD	7.18 ± 1.09 IU/g Hb
Glutathione reductase with FAD	10.4 ± 1.50 IU/g Hb

* For U.S.-European and U.S.-African subjects see p. 73 for other groups.

Enzyme	Activity at 37° C (mean ± standard deviation)
Glyceraldehyde phosphate dehydrogenase	226 ± 41.9 IU/g Hb
GOT without PLP	3.02 ± 0.67 IU/g Hb
GOT with PLP	5.04 ± 0.90 IU/g Hb
Hexokinase	1.16 ± 0.17 IU/g Hb
low S	50.5 ± 5.38%
Lactate dehydrogenase	200 ± 26.5 IU/g Hb
Monophosphoglyceromutase	19.3 ± 3.84 IU/g Hb
low S	49.8 ± 5.48%
NADH-methemoglobin reductase	3.40 ± 0.50 IU/g Hb
Phosphofructokinase	11.01 ± 2.33 IU/g Hb
low S	16.5 ± 2.01%
low S + ADP	29.9 ± 4.15%
Phosphoglucomutase	5.50 ± 0.62 IU/g Hb
Phosphoglycerate kinase	320 ± 36.1 IU/g Hb
low S	56.2 ± 5.33%
Pyruvate kinase	15.0 ± 1.96 IU/g Hb
low S	14.9 ± 3.71%
low S + FDP	43.5 ± 2.46%
6-Phosphogluconate dehydrogenase	8.78 ± 0.78 IU/ Hb
low S	62.4 ± 4.21%
Triose phosphate isomerase	2111 ± 397 IU/g Hb

appendix 2

effect of temperature on red cell enzyme assays

Enzyme	Activity at 30° C / Activity at 37° C (mean ± 1 standard error n = 5)	Activity at 25° C / Activity at 37° C (mean ± 1 standard error n = 5)
Acetylcholinesterase	0.822 ± 0.035	0.730 ± 0.053
Adenosine deaminase	0.750 ± 0.022	0.489 ± 0.021
Adenylate kinase	0.770 ± 0.061	0.553 ± 0.052
low S	1.221 ± 0.097*	1.350 ± 0.147*
Aldolase	0.628 ± 0.051	0.548 ± 0.089
low S	1.057 ± 0.122*	0.932 ± 0.113*
Catalase	0.861 ± 0.015	0.762 ± 0.017
Diphosphoglycermutase	0.710 ± 0.033	0.504 ± 0.038
Enolase	0.700 ± 0.018	0.445 ± 0.010
low S	1.090 ± 0.130*	1.189 ± 0.132*
Epimerase	0.761 ± 0.028	0.644 ± 0.030
Glucose phosphate isomerase	0.760 ± 0.019	0.590 ± 0.014
low S	1.027 ± 0.045*	1.082 ± 0.049*
Glucose-6-phosphate dehydrogenase	0.815 ± 0.031	0.559 ± 0.034
low S	0.937 ± 0.075*	1.118 ± 0.116*
WHO method	0.730 ± 0.035	0.504 ± 0.020
low S	1.095 ± 0.034*	1.153 ± 0.033*
Glutathione peroxidase	0.867 ± 0.013	0.818 ± 0.014
Glutathione reductase without FAD	0.714 ± 0.005	0.543 ± 0.047

Enzyme	$\dfrac{\text{Activity at }30^\circ\text{ C}}{\text{Activity at }37^\circ\text{ C}}$	$\dfrac{\text{Activity at }25^\circ\text{ C}}{\text{Activity at }37^\circ\text{ C}}$
	(mean $\pm$ 1 standard error $n = 5$)	
Glutathione reductase with FAD	0.735 ± 0.009	0.562 ± 0.020
Glyceraldehyde phosphate dehydrogenase	0.699 ± 0.044	0.520 ± 0.013
GOT without PLP	0.867 ± 0.043	0.601 ± 0.019
GOT with PLP	0.789 ± 0.058	0.561 ± 0.015
Hexokinase	0.709 ± 0.037	0.477 ± 0.022
low S	$1.087 \pm 0.052*$	$1.213 \pm 0.076*$
Lactate dehydrogenase	0.670 ± 0.019	0.440 ± 0.009
Monophosphoglyceromutase	0.696 ± 0.015	0.399 ± 0.020
low S	$1.164 \pm 0.025*$	$1.373 \pm 0.036*$
NADH-methemoglobin reductase	0.870 ± 0.021	0.733 ± 0.020
Phosphofructokinase	0.750 ± 0.045	0.580 ± 0.014
low S	$0.873 \pm 0.085*$	$0.902 \pm 0.089*$
low S + ADP	$0.968 \pm 0.103*$	$1.122 \pm 0.088*$
Phosphoglucomutase	0.643 ± 0.011	0.416 ± 0.008
Phosphoglycerate kinase	0.735 ± 0.047	0.604 ± 0.056
low S	$1.065 \pm 0.102*$	$1.059 \pm 0.086*$
Pyruvate kinase	0.689 ± 0.037	0.432 ± 0.026
low S	$0.868 \pm 0.034*$	$1.170 \pm 0.081*$
low S + FDP	$0.976 \pm 0.065*$	$1.063 \pm 0.050*$
6-Phosphogluconic dehydrogenase	0.668 ± 0.013	0.486 ± 0.025
low S	$1.091 \pm 0.082*$	$1.116 \pm 0.036*$
Triose phosphate isomerase	0.656 ± 0.051	0.475 ± 0.047

* Low S values are expressed as ratio of % of the regular system.

appendix 3

levels of intermediates in red cells of normal adults

| Intermediate | Concentration | |
	nmoles/g Hb	nmoles/ml RBC
ADP	635 ± 105	216 ± 36
AMP	62 ± 10	21.1 ± 3.4
ATP (Caucasians)	4230 ± 290	1438 ± 99
ATP (Negroes)	3530 ± 301	1200 ± 102
2,3-DPG	12270 ± 1870	4171 ± 636
GSH	6570 ± 1040	2234 ± 354
GSSG	123 ± 4.5	42 ± 1.53
G-6-P	82 ± 22	27.8 ± 7.5
F-6-P	27 ± 5.8	9.3 ± 2.0
FDP	5.6 ± 1.8	1.9 ± 0.6
DHAP	27.6 ± 8.2	9.4 ± 2.8
3-PGA	132 ± 15.0	44.9 ± 5.1
2-PGA	21.5 ± 7.35	7.3 ± 2.5
PEP	35.9 ± 6.47	12.2 ± 2.2

	μM in whole blood
Lactate	932 ± 211
Pyruvate	53.3 ± 21.5

appendix 4

frequently used molecular weights

Compound	Anhydrous molecular weight (free acid)
Acetylthiocholine iodide	289.2
Adenosine diphosphate (ADP)	427.2
Adenosine monophosphate (AMP)	347.2
Adenosine triphosphate (ATP)	507.2
Aspartic acid	133.1
2,3-Diphosphoglycerate (2,3-DPG)	226.0
5,5'-Dithiobis(2-nitrobenzoic acid) (DTNB)	396.4
EDTA	292.2
Flavin adenine dinucleotide (FAD)	830.0
Fructose-1,6-diP	340.1
Fructose-6-P	260.2
Galactose	180.1
Galactose-1-P	260.2
Glucose	180.1
Glucose-1-P	260.2
Glucose-1,6-diP	340.1
Glucose-6-P	260.2
Glutathione, oxidized (GSSG)	612.7
Glutathione, reduced (GSH)	307.3
D-Glyceraldehyde-3-P (D-GAP)	170.1
KCl	74.5
α-Ketoglutaric acid	168.1
KH_2PO_4	136.1
K_2HPO_4	174.2
$MgCl_2 \bullet 6H_2O$	203.3
Methylene Blue	373.9

Compound	Anhydrous molecular weight (free acid)
NAD	663.4
NADH	665.4
NADP	743.4
NADPH	745.4
NaCl	58.5
Phosphoenolpyruvate (PEP)	168.0
6-Phosphogluconic acid (6-PGA)	276.2
2-Phosphoglyceric acid (2-PGA)	186.1
3-Phosphoglyceric acid (3-PGA)	186.1
Pyridoxal-5-phosphate	247.2
Pyruvic acid	88.0
t-Butyl hydroperoxide	90.1
Tris	120.1
UDPGalactose	556.1
UDPGlucose	556.1

Index and Abbreviations